KV-476-088

THE DEVELOPMENT OF THE CHLORINITY/SALINITY CONCEPT IN OCEANOGRAPHY

FURTHER TITLES IN THIS SERIES

1 J.L. MERO
THE MINERAL RESOURCES OF THE SEA

2 L.M. FOMIN
THE DYNAMIC METHOD IN OCEANOGRAPHY

3 E.J.F. WOOD
MICROBIOLOGY OF OCEANS AND ESTUARIES

4 G. NEUMANN
OCEAN CURRENTS

5 N.G. JERLOV
OPTICAL OCEANOGRAPHY

6 V. VACQUIER
GEOMAGNETISM IN MARINE GEOLOGY

Elsevier Oceanography Series, 7

THE DEVELOPMENT OF THE CHLORINITY/SALINITY CONCEPT IN OCEANOGRAPHY

by

WILLIAM J. WALLACE

Department of Physical Sciences,
San Diego University,
San Diego, Calif., U.S.A.

ELSEVIER SCIENTIFIC PUBLISHING COMPANY *Amsterdam - London - New York 1974*

ELSEVIER SCIENTIFIC PUBLISHING COMPANY
335 Jan van Galenstraat
P.O. Box 211, Amsterdam, The Netherlands

AMERICAN ELSEVIER PUBLISHING COMPANY, INC.
52 Vanderbilt Avenue
New York, New York 11017

Library of Congress Card Number: 72-97440

ISBN 0-444-41118-6

With 3 illustrations and 49 tables.

Printed in The Netherlands

To

Joseph W. MacQuade, Jr.

a fellow laborer in the same vineyards

PREFACE

In 1899 a group known as the Swedish Hydrographic Commission prevailed upon the King of Sweden to convene a meeting of the world's oceanographers to discuss problems vexing sea study at that time. Foremost amongst these was the density problem. The density of sea water is a function of temperature, salinity and pressure and is, in itself, calculated from known relationships of these three parameters used largely in connection with geostrophic computations.But the problem at that time was specifically that of salinity (and chlorinity) determination. There didn't seem to be any question about the validity of the theory that the ratio of the major ionic constituents in sea water was a constant. "Normal science is predicated on the assumption that the scientific community knows what the world is like" (T.S. Kuhn, *The Structure of Scientific Revolutions*, Phoenix, Chicago, 1964).

To determine salinity the (S‰)/(Cl‰) ratio was best used, since chlorine is the most abundant ion in sea water and is rapidly, accurately and easily determined. There was, however, definite question as to the value of the coefficient (of chlorine) obtained [(S‰)/(Cl‰) = k = coefficient of chlorine)]. Forchhammer, the accepted originator of the coefficient of chlorine (as well as the term "salinity") had arrived at 1.812 in 1865. Dittmar, analyzing the 77 water samples of the HMS "Challenger" expedition, has arrived at a coefficient of 1.8058. Of particular concern was the fact that there was no actual accepted definition of salinity per se but only as a determined quantity (i.e., as by the chlorine determination and use of a coefficient). In other words, the accepted paradigm was in a state of crisis. There was need of some definition. The situation was not one of radical change or scientific revolution. No scientific theory is ever discarded unless there is another alternate, and there was in 1899 no other available. The situation was rather that of normal scientific activity; namely, the bringing of theory and fact into closer agreement.

The scientist charged by the 1899 conference to resolve the problem was Martin Knudsen. By 1902 his group had a gravimetric definition of salinity, a coefficient of chlorine, an equation relating the two (S‰ = 1.805 Cl‰ + 0.030), and the famous *Hydrographical Tables* as well.

Years ago when I first encountered the Knudsen equation, as it is most often called, I wondered about the use of the constant 0.030. It was immediately bothersome to me since it violated the notion that the equation was one of state which it was supposed to be, as well as the fact that it made the ratio of ions (or S‰ to Cl‰) upon which the equation is based not constant. A lengthy study of all of the endeavors of Knudsen and his team disclosed the fact that Knudsen was well aware of these problems. What Knudsen said and the questions he raised in the lengthy published reports people did not read. Only the *Hydrographical Tables* were ever read and used. But the marine community was primarily looking for standardization – so everyone could play the game with the same rules.

Man's ideas about the physical world have changed with time. Not only have specific theories replaced others only to be themselves replaced, but some fundamental concepts also seem to have changed. Lucretius, for example, would have said that nothing happens without a material or physical cause. Or, in other words, one only attained true knowledge when all things came to pass without the intervention of the gods. Modern science, however, says that everything happens unless there is a reason for it not to happen (i.e., the conservation laws).

What I have tried to do in this book is trace man's ideas about the sea's saltness from antiquity to the present time and show with the salinity/chlorinity concept something of the way science operates. In so doing this work also becomes, at least until 1902, a rather detailed history of marine chemistry.

The study of science as well as its history and historiography are fascinating. It is, I think, unfortunate that it is generally only those who are engaged in these studies that tend to see science as a human activity – probably the most human of activities. I would hope that works such as this one help indicate this.

ACKNOWLEDGEMENTS

A work of this type probably necessitates more assistance from others than most. Although it would be very difficult to mention all those who have been of help since it was begun seven years ago, it would be remiss not to cite a number of these people.

I would like to thank Dr. Marie Boas Hall of the Imperial College of Science and Technology in London, and Dr. Allen G. Debus of the University of Chicago for their comments on Robert Boyle and the silver nitrate test. Dr. Edward D. Goldberg of the Scripps Institution of Oceanography and John P. Riley of the University of Liverpool were of much assistance in early bibliographic questions.

Dr. Jorgen Pindborg, Dean of the University of Copenhagen Dental

School. not only was most helpful with translations from the Danish but also in supplying at least one especially important paper by Forchhammer from the University's library. Dr. Robert Stadler of Heidelberg was of particular assistance in the identification of compounds from German papers on sea water analysis written in the first half of the nineteenth century.

This treatise involved the use of many libraries. The friendly assistance afforded, for example, by the great Widner Library of Harvard as well as the Library of the Museum of Comparative Zoology with the granting of visiting scholar status which involved desk and office space, complete admittance to the stacks, as well as faculty privileges at no cost, was tremendous.

In closing, I would like to thank Dr. Robert J. Morris, who waded through a number of often barely legible drafts of this weighty tome and in whose hands the book improved measurably, and I would like to offer a special vote of thanks to Miss Audrey Lee Krueger for her kindness in the deciphering and transposal of the first good draft from handwritten scrawl to typed copy. My good friend Dr. Thomas H. Foote, of the Evergreen State College in Olympia, Washington, was most generous with support, comments and his assistance in the reading and proofing of the final draft.

CONTENTS

CHAPTER 1

THE VIEWS OF ANTIQUITY ON SALT AND SEA WATER

WATER ANALYSIS PRIOR TO BOYLE

In 1674 the famous English chemist and natural philosopher Robert Boyle (1627–1691) published a treatise entitled *Observations and Experiments on the Saltness of the Sea*. This was an extraordinary work which in the opinion of several (modern) writers established him as the founder of the science that is now referred to as chemical oceanography (258*, p.5; 284*, p.109). It was the first truly definitive, scientific, quantitative paper dealing with the sea and its various parameters such as temperature, depth, and salt content. Prior to this work by Boyle, there had been little scientific interest in the sea in general, although there had been some interest in the salt it contains.

Salt has always been a commodity highly prized by man and as such has often figured in his affairs. It was mentioned often in ancient writings. A number of customs and superstitions arose concerning salt and its use. It was considered the height of hospitality to offer salt to a traveler or visitor (284, p.109). Salt was often used as a medium of exchange. The soldiers of the Roman Legions were often paid in salt – hence the word salary (284, p.109). This custom of remuneration with salt occurred in modern times when the Italian army, having conquered Ethiopia in 1937, tried to hire native workers in the more remote regions of that country. These people refused the Italian lira but were readily secured when offered, for pay, disks made of salt (284, p.109).

The sea is a dilute solution of chemicals. When sea water is evaporated the primary resultant compound is salt, a fact probably known for almost as long as sea water itself has been known. Salt was considered to be the only constituent of sea water other than the water itself for most of recorded history or until the seventeenth century when it began to be known that additional compounds called salts were present.

The oldest of chemical industries – the production of salt from the sea – was practiced in a number of crude forms by the people who dwelt near the sea. The methods these people employed varied somewhat. In arid or semi-arid regions it was simply a question of allowing sea water to evaporate

* Numbers refer to those of the bibliography (pp. 209–220).

naturally once it had been put into basins. In more temperate regions evaporation was not a feasible method. From prehistoric times the Scandinavian peoples separated salt by allowing the sea water to freeze (284, p.109). The ice was removed and the resulting brine concentrate was thrown on charcoal fires and the salt crystals removed. This method was essentially the same as that used by the ancient Britons, as mentioned by Julius Caesar in his writings (284, p.109).[1]

These methods were developed probably by trial and error primarily from the need for salt. Although it would have to be true that these men developed a feel as to how much salt could be gleaned from the evaporation of a certain amount of sea water, this was, depending on the method used, only a rough idea. A quantitative and qualitative estimate of the contents of sea water was impossible until scales and chemistry had advanced far enough at least to the point where true compounds and elements were capable of being recognized.

Since salt and the sea have always been important to man it comes as no surprise that the ancients would have become interested in and speculated about the sea, its saltness, and the origin of both. The earliest known writings that exist that treat these topics are Greek. Among these, the writings of Aristotle loom particularly large because they are the primary source of not only his own ideas on this subject but also those of his predecessors although he often did not mention them by name.

Aristotle (384–322 B.C.) was interested in the explanation of all nature. Greece has largely been a seafaring nation, and Aristotle's comments on the sea and its contents, especially those on marine life, are numerous.[2]

Aristotle did describe the physical characteristics of a number of saline bodies of water – the Dead Sea, Red Sea, Black Sea, Aegean Sea, as well as the Mediterranean (5, vol.7: 393ᵃ29–393ᵇ2) – and it is clear that he was aware of the existence of the Atlantic (5, vol.7: 392ᵇ22–393ᵃ17) which he regarded as the greatest ocean though he did consider all the oceans as one since they were continuous (5, vol.2, book II: 297ᵃ9–16).

In a number of places throughout his works Aristotle commented on salt and it is evident that his knowledge on the physical and chemical properties of this substance was as good if not better than that of anyone else for almost two thousand years to come. Several methods for the preparation of salt were mentioned by him. For example, in (and around) the city of Utica, in Libya, holes were dug to a depth of 18 ft. whereupon the salt appeared white and not solid. Once the gummy material was brought up and subjected to the sunlight it became hard (5, vol.6: 844ᵃ7–16). Aristotle told of a spring in the country of the Illyrians whose water in the spring when it ran was collected and dried five or six days and yielded excellent salt (5, vol.6: 824ᵇ9–23). He further stated that it was necessary that these people pro-

[1]For Notes, see pp. 193–207.

cure salt in this fashion as they were some distance from the sea and that they needed it for their life and that of their cattle.

Salt was water-soluble and would dissolve in oil (5, vol.3: 383^b 12–17). He also knew of other similar substances other than salt itself – and he mentioned often, for example, natron (sodium carbonate). The noise that salt makes when thrown on fire was explained by Aristotle as being due to the presence of small quantities of moisture in the salt which when heated evaporated and in bursting out rent the salt (5, vol.3: 902^a 1-4–904^a 4-15).[3] Aristotle also believed that there was some difference in sea salt and salt. The salt from the sea (prepared by evaporation) was weaker in saltiness and was generally not as white and less lumpy than "normal" salt (5, vol.6: 399^a 30–35).[4] Aristotle presented most of what was probably the previous ideas as to the origins of the sea and its saltness.[5] Much of this presentation of the views of others was undoubtedly given just for the sake of information. Yet most were probably presented, as was common with Aristotle, simply to show that he believed that these opinions could not be an accepted or correct explanation.

The water which forms the sea was said to come from the springs that feed it (5, vol.3: 353^b 33–35). Another philosopher said that all the earth was at first surrounded with moisture, some of which later formed the sea (attributed to Anaximander and Diogenes of Appolonia) (5, vol.3: 353^b 5–12) in a process of drying which would ultimately end in the loss of the sea. A third likened the sea to a kind of sweat exuded by the earth by the action of the sun (attributed to Empedocles) (5, vol.3: 353^b 11–13) which also nicely explained the sea's saltness since sweat is salty.[6] Another philosopher simply attributed the saltness to the earth that the water picked up as it came in contact with the earth as it ran over it, just as water strained through ashes is known to be salty (attributed to Metrodorus of Chios and Anaxagoras). The sea itself was explained by the accumulation of the run-off (5, vol.3: 353^b 13–16).

Aristotle began stating his own ideas by saying that all of his predecessors were prone to consider the sea as the origin and source of moisture and all water. He states further that since this is so there is some need to explain why rivers which originate from the sea even though they flow back to it are not salty but sweet (5, vol.3: 354^b 15–23):

> Now the sun, moving as it does, sets up processes of change and becoming and decay, and by its agency the finest and sweetest water is every day carried up and is dissolved into vapour and rises to the upper region, where it is condensed again by the cold and so returns to the earth. This, as we have said before, is the regular course of nature.
>
> (5, vol.3: 354^b 26–30).

In explaining this, Aristotle returned to his idea of natural place and said that the sea is the filling of low places due to the action of rivers, and the condensation of moisture (rain) is not the natural place of the sea but simply that of water (5, vol.3: 355^{b} 1–3).

> Hence all rivers and all the water that is generated flow into it (the natural place): for water flows into the deepest place, and the deepest part of the earth is filled by the sea. Only all the light and sweet part of it is quickly carried off by the sun, while the rest remains for the reason we have explained. (5, vol.3: 355^{b} 15–20)

The sweet water is lighter than the salt water which is heavy and so is carried off by the heating action of the sun. The rest remains.[7] If, however, one maintains that an admixture of earth made the sea salty and that the rivers in washing the earth pick up a variety of flavors which by this mixture make the sea salty, then it would follow that rivers are also salty. But, Aristotle points out, they are not (5, vol.3: 357^{a} 14–24). Yet it was clear to Aristotle that the salt in the sea must be due to the mixture of something.

> To say that it was burnt earth is absurd; but to say that it was something like burnt earth is true. We must suppose that just as in the cases we have described, so in the world as a whole, everything that grows and is naturally generated always leaves an undigested residue, like that of things burnt, consisting of this sort of earth. All the earthy stuff in the dry exhalation is of this nature, and it is the dry exhalation which accounts for its great quantity. Now since, as we have said, the moist and the dry evaporations are mixed, some quantity of this stuff must always be included in the clouds and the water that are formed by condensation, and must redescent to the earth in rain. This process must always go on with such regularity as the sublunary world admits of, and it is the answer to the question how the sea comes to be salt. (5, vol.3: 358^{a} 15–28)

This attributes the saltiness to the earthy residue in dry exhalations and the recycling of this residue and consequent build-up in the sea.[8] This earthy residue is probably earth or akin to earth or "particles" thereof in the air which the rain brings down. In certain times of the year the rain must be heavier (as in autumn) (5, vol.3: 378^{b} 4–8) and descends not only the quickest but in greater amount. Although quite familiar with salt as a substance he never really explained what it was (in sea water) but only vaguely how it got there. Specifically it appears that salt in salt water is somehow due to the earth but more so that there must be something else other than simply a mixture of earth and water. If sea water was purely earth and water then the earth should seek its natural place below that of water and sink, therefore rendering salt water fresh. In fact this is borne out in the case of fresh water (5, vol.4: 283^{b} 26–28) as he states that earth is in its natural state in fresh water and later accounts for alluvian formation and sedimentation at river mouths. There is, then, an earthy quality about the salt in sea water but beyond this its nature is vague.

The sea, according to Aristotle, has a continual tendency to become more salty and this is due to the action of heat. Some of the saltness (from the earth – not from the sea) is constantly being drawn up with the fresh water. But salt water when it is evaporated yields a fresh (sweet) vapor and when condensed into the liquid again does not reform salt water. These two processes counteract one another, and the sea's saltness, then, remains constant on the whole (5, vol.3: 358^{a} 12–24).

Sea water was hotter than fresh water as it was according to Aristotle both wet and dry whereas fresh water was cold and wet. This was a bit disturbing in terms of the diagram of the four elements and their properties of hot, cold, dry, wet (Fig. 1), since all physically existing substances contained all four "elements". Fresh water has a preponderance of "elemental" water

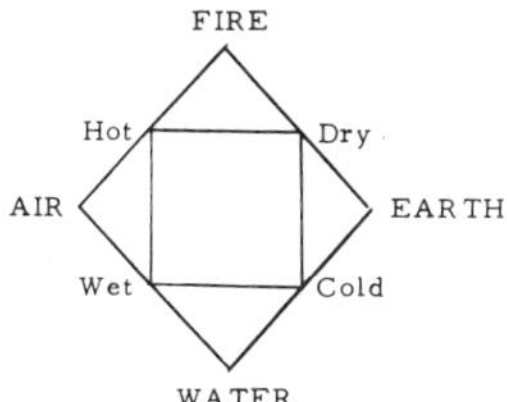

Fig. 1. The four Elements in association with the four Qualities (273, p.8).

whereas sea water contains more "earth" or earth-like material. There is little to indicate that Aristotle used the diagonal in this diagram. Sea water, then, would have some properties of earth. The heat (apparently from the sun and the earth in that order) can take hold of the salt water and rarify its parts (5, vol.4: 824^{b} 26–30). The sun (i.e., its heat) can more easily attract the fresh water because it is lighter and nearer the surface and thereby removes the fresh water leaving the water at the surface saltier (5, vol.7: 934^{a} 24–33). For this reason, it was said that water of wells is saltier at the upper regions and that the upper parts of the sea also contain more salt than the depths. There is a contradiction here. According to Aristotle fresh water is lighter than salt water but that the surface regions due to the evaporation of the fresh water are saltier and therefore more heavy. It would seem that the heavier surface water, having become saltier due to the sun's action and its own innate heat, should sink to a more natural place. Nowhere, however, is this problem either really mentioned or treated or much less solved.

Aristotle mentioned a number of times that salt water was heavier and more dense than fresh water (5, vol.3: $355^{a}33$, $359^{a}5$), and salt water would seek a lower level (5, vol.3: $824^{a}28$–30). He cites an experiment for density – an egg floats in salt water but not in fresh water (5, vol.3: $824^{a}20$). In fact, he mentioned that the difference in consistency of fresh water and salt water accounted for the sinking and near-sinking of a number of freighting vessels upon leaving the sea and entering rivers (5, vol.3: $359^{a}6$–11).[9]

One of Aristotle's experimental proofs that saltness is due to the admixture of some substance was with the use of a completely closed wax vessel. This container was lowered into the sea and:

> . . . then the water that percolates through the wax sides of the vessel is sweet, the earthy stuff, the admixture of which makes the water salt, being separated off as it were by a filter.[10]

As previously mentioned (p.4) (5, vol.3: $359^{a}2$–5) Aristotle was aware that the sea contained other than just salt and commented on both its salt and bitter taste (5, vol.3: $354^{b}1$–2) and wondered as to the cause. Sea water when concentrated as in the case of the Red Sea, for example, became increasingly more salty and bitter. He likened the bitter taste of sea water to the bitter and salty quality of urine but explained it, again, as the admixture of something earthy with the water (5, vol.3: $357^{b}1$–8). He added: "Why is the sea salty and bitter? Is it because the juices in the sea are numerous? For saltness and bitterness appear at the same time" (5, vol.7: $935^{a}35$–37). This is important and it should be emphasized. Aristotle apparently was the first person to have noticed and attempted to explain the bitter quality of sea water – a point which did not occur in the literature for at least two thousand years later. More specifically he was the first to mention something other than salt and water in sea water. He also, strangely enough, attributed a fatty quality to sea water and said that in hot weather especially a fatty oily substance formed on the surface due to its lightness (which, by the way, gave water its transparency) having been excreted by the sea which has fat in it (5, vol.7: $932^{b}16$–24).[11]

Aristotle's attempts to explain the saltness of the sea were hardly altogether clear and the wax container experiment is just one simple example. Later confusion and misconceptions did arise concerning this subject area. From a compositional standpoint, Aristotle tried to answer the questions: why sea water was salty, why water which is naturally fresh became salt, and what was the nature of material that caused the bitter taste present in sea water. While the answers which are given above are often vague and incomplete, especially, for example, with his answer to the question of the cause of the bitterness, several points should be made here. It is especially important to note that Aristotle did not attribute the saltness of the sea to the action of the sun per se. Rather the heat of the sun was only one of the possible causes of the saltness – but it was the major one. It is quite certain that he believed that the basis of all waters, including the sea, was fresh water and saltness was only accidental (5, vol.3: 824^{a}4–7).

Another important and influential work of antiquity was that of Pliny. In 77 A.D., Pliny the Elder (Gaius Plinius Secundus, 23–79 A.D.) completed the encyclopedial *Natural History* (253). This set of 37 books dealt with almost every imaginable topic. Included in its realm were topics in astronomy, meteorology, geography, mineralogy, zoology and botany and related subjects. There were numerous remarks on salt and sea water.

Pliny was quite familiar with salt and salts. He mentioned that "a civilized life is impossible without salt, so necessary is this basic substance" (253, Book 31, p.433). In Book 31, which was supposed to deal with the remedies derived from aquatic animals, he treated at length of waters and salt. Salt for Pliny was divided into two general classes: natural salt and artificial – both formed in several ways. The distinction was essentially this. Natural salt is that which is found in large earthly deposits as might be dug or quarried (253, Book 31, p.425). Artificial salt, however, is that formed by the evaporation of some water, the number of types of this artificial salt apparently being determined by the type of water that is dried – 'such as sea, well, spring water – the most plentiful artificial salt being made from evaporation of sea water. Fresh water, however, must be added to and mixed with the sea water for the salt to be deposited (253, Book 2, p.301).[12] The salt formed from the pouring of salt water onto burning logs as is practiced in provinces of France and Germany would be then a specific type of salt. The wood even turns into salt when brine is poured onto it (253, Book 31, pp.427–429).[13] These artificial salts can have virtually any number of flavors and/or colors, and since they were formed only by this process of evaporation from almost any natural waters it is evident that these salts were not all sea salt. Pliny, however, knew of other salts. By name he mentioned soda (Latin equivalent, 253, Book 31, p.443), which was made like salt (by evaporation) and while it is natural, it was in Egypt made artificially (253, Book 31, p.445).[14]

Salt was by nature firey but was hostile to fire and corroded all things.[15] It had an appreciable number of uses in medicine which he at length enumerated (253, Book 31, XLIV. p.439) and he quoted also at length the medical uses of salt in fresh water (253, Book 31, pp.437–439) as well as sea water. He listed a number of different types of mineral waters, such as sulfur, bitumen and soda (253, Book 31, p.413).

Pliny appears to have been the first person to give an early quantitative estimate for the amount of salt in sea water by which one could make sea water:

> . . . if more than a sextarius of salt is dropped into four sextari of water,

the water is overpowered, and the salt does not dissolve. However, a sextarius of salt and four sextari of water give the strength and properties of the saltest sea. But it is thought that the most reasonable proportion is to compound the measure of water given above with eight cyathi of salt. This mixture warms the sinews without chafing the skin.
(253, Book 35, p.421) [16]

Fresh water could be separated from sea water a number of ways. Fleeces could be spread about the deck of a ship and once moist by the absorption of evaporated sea water, wrung out to produce fresh water. Hollow wax balls or empty vessels with their mouths sealed once immersed in the sea would fill with fresh water.[17] On land the sea water could be filtered through clay to produce the fresh.[18]

Pliny's explanation of the sea's saltness was:

Consequently liquid is dried by the heat of the sun, and we are taught that this is the male star, which scorches and sucks up everything; and that in this way the flavour of salt is boiled into the wide expanse of the sea, either because the sweet and liquid, which is easily attracted by fiery force, is drawn out of it, but all the harsher and denser portion is left (this being why in a calm sea the water at a depth is sweeter than that at the top, this being the truer explanation of its harsh flavour, rather than because the sea is the ceaseless perspiration of the land), or because a great deal of warmth from the dry is mixed with it, or because the nature of the earth stains the waters as if they were drugged. (253, Book 2, pp.349–351)

This explanation is not altogether Aristotelian. Here only the sun's "fiery force" is responsible for the saltiness. As in the case of Aristotle, Pliny thought the salt content should be greater at the surface due to the loss of water here. Yet along with Aristotle Pliny knew that salt water was more dense than fresh water (253, Book 2, p.353), and he indicates that patches of fresh water can be found floating on the surface of the sea (253, Book 2, p.353). Like Aristotle he does not reconcile the contradiction. Interesting, however, is the addition of the condition of a calm sea. This was not mentioned in Aristotle and it seems to indicate some idea of mixing (oceanic). The ceaseless perspiration would appear to indicate that the seas should be constantly increasing in size; this, however, was never set forth. Pliny said that a number of springs could be found in the sea and that in a great many places fresh water could be drawn from the sea (Book 2, p.355). This is in disagreement with Aristotle. The saltness of the sea varied with the season, the sea being saltier in autumn (Book 2, p.361) (as well as warmer in winter).

There is no reason to believe that the notions presented by Pliny were not indicative of the thoughts on these subjects for his time.

Another Roman and a contemporary of Pliny was the philosopher Seneca (3 B.C.–65 A.D.). Seneca's views as to the nature of the world appeared in his *Quaestiones Naturales*. Seneca was a keen observer, and much of the *Quaestiones Naturales* present his own observations with more originality than, for example, Pliny. Seneca had noticed that the water level and the salinity of the sea remained constant even though water was constantly being added by rivers and rain. The constancy was, he believed, due to the evaporation of the sea's waters. He believed that saline waters could be filtered by earth and attributed the formation of calcareous tuffs to this action. Seneca thought that the world in the beginning was characterized by a primordial ocean, and the substances dissolved therein separated out over some space of time. He was unique in his classification of the waters of the earth and divided them into four categories.

> (*i*) oceanic waters which have existed since the beginning of time; they constitute the bulk of the water on the earth and are the source from which all others are derived; (*ii*) subterranean waters which circulate in the earth and soil and which emerge at the surface as springs; (*iii*) waters which circulate or remain stagnant on top of the soil; (*iv*) water present in the atmosphere. (258, p.3)

Erosion by waters (streams, currents, waves) and their transport, and sedimentation at the mouths of rivers to form deltas were mentioned by Seneca. Although he knew that solubility of a substance was in some way related to the water's temperature and that the temperature of the sea varied, he seems to have believed that the ocean's saltness was a constant (258, p.3).

These views present almost entirely the theories and observations of the ancient philosophers with regards to the sea and its saltness. And it is essentially these views that prevailed and were available to the Western World over one thousand years later. Some of the ideas concerning the sea and its salt by the time of the reintroduction of Greek knowledge into Western Europe had been modified and added to by the Arabian "Empire".

Since the Arabian ships plied the seas for some time it is not surprising that Arab philosophers should become interested in the sea and its geography. One of these men, who appears to have written the most extensively on the subject, was El Mas 'Ūdi (fl. 915 A.D.). Mas 'Udi's book entitled *Meadows of Gold and Mines of Gems* (86) is in reality an historical encyclopedia. This work is much like Pliny's *Natural History* but more closely followed Aristotelian ideas although heavily seasoned with Arabic mythology and explanations. While it is often difficult to ascribe specific points on theories to specific people, Mas 'Udi, in general, was better about his work than Pliny or Aristotle on this point. Volume I (of the translation) treats primarily of the sea and related topics. The physical description of the seas fills three chapters.[19] Springs, rivers and all waters have their ultimate source in the greatest sea (86, p.230). The identity of the greatest sea is somewhat obscure but it is clear that he does not mean the Atlantic which he does mention by name (Okianos), although he added that (86, p.282) there were some people who thought that the Atlantic was the origin of all other seas and waters.

He divided water into three classes, the criteria for categorizing being with regard to its ebb and flow. The ebb and flow could be: (*1*) apparent and evident; (*2*) not apparent and evident; (*3*) no ebb and flow at all. The ebb and flow was caused and influenced by the moon (86, pp.272–273). The motion of the sea coincided with the course of the winds, according to Mas 'Udi (86, p.278).[20]

In discussing the cause of the sea's saltness, Mas 'Udi wrote:

> The water which flows into the sea from the high and low grounds of the earth absorbs, according to its nature, the salt which the earth throws out into its basin; the particles of heat which emanate from the sun and moon, and cause the water, being mixed with it, to come forth from the earth, raise and evaporate the water by their raising (expansive) power, the finer particles of water, when it is above, are turned into rain. This process is constantly repeated, because this water becomes again salt; for the earth imbues it again with saline particles, and the sun and moon deprive (the sea) again of the finer and sweet portions of the water (by evaporation). It is for this reason that the sea remains unchanged both in quantity and (specific) weight (salt dissolved in it); for the heat raises the finer portion of the sea water, and changes it into atmospheric humidity, in the same proportion as the same water flows again into the sea, in the form of streams, after it has become terrestrial humidity; for, being in the form of streams, it has a tendency to stagnate, and to form marshes flowing to the deepest places of the earth, and so it comes into the bed of the sea. The quantity of water remains, therefore, constant and is neither increased nor diminished.
>
> (86, pp.302–304)

The Aristotle influence is obvious here. It should be noted that Mas 'Udi has attributed as cause of saltness the result of the heating or complete burning process. In so doing he omitted Aristotle's full answer to the question. Mas 'Udi did say, however, that if the sea does remain quiet for some time it will become saltier, heavier and denser. The loss of the water due to evaporation is obviously the cause. Also, according to Mas 'Udi the sea remains constant in volume and saltness. Arabic oceanic thought placed great emphasis on the moon which was not so in the Greek thinking, especially with respect to possible heating processes.

The heat is communicated to the atmosphere and hence the oceans by the moon. This heat also expands the water which increases in volume, and causes the spring and monthly tides (86, p.274). The sea bottom becomes warm[21] and by heating produces sweet water at depth (86, p.273) which is then changed into sea water and becomes warm, as happens in cisterns and wells. This segment is not clear. While Mas 'Udi said that salt water is produced at the sea bottom as well as the surface, he implied that the surface waters should be more salty (probably due to greater heat at the surface). Again, as with Aristotle and Piny, Mas 'Udi knew that sea water was heavier than fresh water (86, p.305). This apparent contradiction remained unsolved. Mas 'Udi also repeated the story of the wax vessel.[22]

This, then represents essentially the knowledge of the sea and especially of its saltness that became available to the Western World by 1200 A.D. Although the Arabs added a number of points about this subject, especially with regard to sedimentation, the thinking was primarily that of the Greeks and especially that of Aristotle in origin. There appears to have been little written by Arabic writers to challenge the traditional Greek masters. There was, however, some tendency to misread Aristotle, especially in his causes for saltness as being due solely to the sun and its action.

The subject of salt was an important one in Arabic writing and there were numerous references to it in the alchemical writings, especially in the later periods. They did enumerate a number of different salts not known to the Greeks such as ammonium salts (ammonium nitrate, ammonium chloride, etc. – 176, p.65). The salt mentioned generally was not sea salt but rather salt in the general sense. It became one of the three Arabian "principles" of metals along with mercury and sulfur.

SOLUTION ANALYSIS PRIOR TO BOYLE

It was suggested over 100 years ago (170, vol.2, pp.50–55) that the procedures for solution analysis were derived primarily from the study of mineral waters. These studies of mineral waters appear to date approximately from the end of the sixteenth century (170, vol.2, p.51). Since it is not overly clear how much of the chemistry of mineral waters was due to Boyle (23, p.3) some brief background on solution analysis prior to Boyle is given here.

Natural waters from time immemorial have been considered to be of medicinal value and possess curative powers. During the Renaissance there developed interest in spa waters (75, p.43). This was primarily due to the studies of the content of local waters by Italian physicians during the Middle Ages. Probably the earliest of such written accounts was a text by Peter of

Eboli, dated 1195 (75, p.43). Another of 1256 by Alderbrandino of Sienna was interesting in that the author made artificial mineral water by boiling water with sulfur. Virtually all of these involved a simple evaporation of the water and the subsequent testing of the residue with the senses, primarily taste. Giacono de Dondi (1289–1359), a medical lecturer at Padua, had suggested testing the residue with red-hot coals so as to notice any hidden odors (75, p.43). In a manuscript dated 1399, Francis of Sienna mentioned the analysis of some waters by distillation and subsequent finding of alum, sulfur and iron in the residue (283, pp.42–43).

The distillation process, at least in principle, is as old as the time of Aristotle. While it played an important part in the history of alchemy (223, p.329) it had virtually little or no role in water analyses of the thirteenth and fourteenth centuries.[23] Distillation by the middle of the fifteenth century was an accepted laboratory practice and there is a wealth of examples of its use in the preparation of medicines at this time (223, p.333). The use of distillation in water analyses from this time was a common practice. Bartolomeo Montagnan (d. c.1460), a professor of medicine at Padua, analyzed the waters of seven springs by initially distilling the water, followed by an examination of the residue in terms of color and taste (75, p.44).

Another professor at Padua, Michael Savonarola (c. 1390–1462) wrote a treatise entitled *De balneis et termis naturalibus omnibus ytalie* (286, vol.4, pp.211–212) which was the most advanced text yet written in regard to solution analysis. Its influence may be traced for over a century (75, p.44). Savonarola regarded tests such as taste to be useful but often not positive enough. He suggested additional tests especially the placing of a portion of the residue into a flame. From this he noted changes in odor or color of the residue as well as changes in the color of the flame. He knew, for example, that sulfur burned with a green flame (75, p.44). Niter and salt he distinguished by the crackling noise the salt emits upon strong heating.

The works of Savonarola spread well beyond Italy. Savonarola's ideas are reflected, for example, in the works of the two Germans, Johann Widman (1513) and Johann Dryander (1535) (75, p.45). The famed anatomist of Padua, Gabriel Fallopius (1523–1563) who was a physician associated with the medical school at Padua who wrote on solution analyses was also greatly influenced by Savonarola. Fallopius' work is important in that he gave a much more detailed procedure for water analysis. The *Tractatus de balneis thermarum ferinarum* (75, p.46) itself was influential and represents the culmination of the Paduan tradition of water analysis (75, p.47). Fallopius was able to recommend distillation followed by sense tests of the residue, drying the solids by the sun on a smooth table so as to be able to see any bright crystals. In this way crystals such as alum, nitre, salt and sulfur could be identified (75, p.47). This was followed by the placing of the substances on a red-hot iron. He suggested the addition of acids (aqua fortis) and subsequent distillation for the identification of metals – copper, for example – by the green substance remaining.

By the sixteenth century the ancient Greeks' texts on medicine, recovered by the fourteenth century, coupled with the move toward the dissection of the human body and the corresponding rise of the medical schools in Italy, created an atmosphere conductive to the pursuit of all things medical. At this time there arose a separate entity, iatrochemistry – or what now might be called "medical chemistry". This was characterized and enhanced, but not by any means begun, by Paracelsus (c. 1493–1541) (223, p.336). It is not surprising that the followers of Paracelsus were interested in mineral water analysis as Paracelsus himself was a medical chemist and laid great stress on the possible and actual medicinal value of natural waters. The German physician and Paracelsian Leonhart Thurneisser zum Thurn (1530–1595) in his *Pison* (1572), for example, introduced a quantitative approach to water analyses. Beside the quantitative quality of his work he also employed solubility tests, some flame tests and crystal observations (240, vol.2, p.150). Thurneisser used in any water analysis a measured, weighed amount of the water filtered into a clear container, distilled to dryness and the residue weighed. This was followed by subsequent tests for specific salts with special attention given to the shapes of the crystal forms of the recrystallized salts. Toward the end of the sixteenth century one of the founders of qualitative analysis (240, vol.2, p.264), Andreas Libavius (1540?–1616), also placed emphasis on the quantitative approach. It was his opinion that mineral waters (those waters containing minerals) were heavier than pure waters.[24] Libavius' work is important because he stressed the value of recrystallization after distillation since specific salts could be separated – for example, salt from saltpetre, alum from atrament. Unlike Thurneisser, Libavius used the oak gall test and did not use the distillate for further indication of dissolved minerals (75, p.50).[25] Paracelsus had described the oak gall test in 1520 (75, p.45) but never indicated any definite analytical procedure.

A host of information from a number of sources indicates that the methods of the mineralogists and metallurgists were similar to the medical analysts in their techniques analysis. This Vanoccio Biringuccio (1480–1539) and Agricola show in their works (though not intentionally) (240, vol.2, pp.33, 45–53).

There were a number of early seventeenth century English analysts of mineral waters. The earliest of these analysts was a minister-physician schooled at Bologna or Ferrara, Walter Bailey, a physician to Queen Elizabeth (75, p.52). He gave an analysis of these waters such as the scheme had appeared in the works of Fallopius (9, p.10). Aside from containing the earliest written English reference to Paracelsus (75, p.50) and references to most of the above mentioned men, his works contain little of his own analyses.

The most comprehensive and original of the English analysts prior to Robert Boyle was Edward Jorden (1569–1632), who received his M.D. at Padua, a member of the Royal College of Physicians and one of the King's physicians. He was evidently a friend of Libavius and was aware of the latest advances in science. In his *A Discourse of Naturall Bathes and Minerall Waters* (1632) he rejected the four Aristotelian elements as the material from which all matter was composed (156, pp.74–78). Reliance purely on distillation techniques with subsequent tasting he regarded as inadequate. He was familiar with virtually all of the authors mentioned above and specifically praised the analytical procedures of Agricola, Fallopius, Baccius and Libavius. He divided the materials in mineral water into eight categories: simple earth, stone, bitumen, salts, mineral spirits, mean metals and spiritual substances (156, pp.17–76). He classified the salts as four types: niter, salt, alum and vitriol and their respective species, "diverse species" (156, p.29). Jorden placed extreme importance on the identification of salts by their crystal form. The specific components remaining as the residue after distillation and separation by recrystallization, according to his suggestion, could be tested for purity by crystallization – saltpeter, for example, would only form needles when pure (156, p.38). Jorden also suggested the use of precipitation as a swifter way to obtain salts without the wait involved in normal crystallization methods, and emphasized precipitation as an analytical procedure.

THE LATE SEVENTEENTH CENTURY

Between the years of the re-introduction of Greek science in Western Europe and the late seventeenth century there had been surprisingly little reference to the sea, especially its saltness. Although the sea was a perennial problem that had bothered natural philosophers for centuries (269, p.159), few authors discussed it extensively. With a few exceptions such as Joan de Jaraua who, in 1546, provided an ingenious variance to the percolation theory by suggesting that sea water was distilled up through mountains (269, p.159), virtually nothing existed in theory or description that was not Greek in origin. Leonardo da Vinci, for example (189, vol.2, pp.63–65; 240, vol.2, p.1), had thought that the springs of the earth dissolved the salts out and carried them to the sea. Salt he said was also extracted by water from ashes and burnt refuse. Julius Caesar Scalinger (240, vol.2, p.10) held a similar view (1557) but had said that he had found by experiment, in taking samples at depth, sea waters to be fresh at depth. Henry Oldenburg (236) (1615?–1677), the first secretary to the Royal Society (123, p.191), mentioned that the only sure way to test the saltness of the sea with respect to its entire profile would be by trial. Further, he stated that even if the sea were found to be fresh at some place at the bottom that it might simply mean it was nothing but a sweet water spring gushing forth at that place. One paper contained a discourse about the use of the mercury hydrometer[26] to measure relative salt content. In this he stated that contrary to some opinions the color of the sea was not related to the degree of saltness. Further the author said as a result of his measurement that the sea was saltier only near the tropics.[27]

Although there was little of a theoretical chemical nature that went beyond that of the Greek knowledge, chemical techniques had improved considerably since Aristotle. In fact, there was a *chemistry* in the seventeenth century. By the middle of the fifteenth century distillation had become a common technique. By mid-seventeenth century there existed not only the common weighing practices but also specific gravity comparisons: a handful of flame tests; a number of specific qualitative tests such as that of oak gall; crystallization; the separation of a number of salts such as alum, sea salt, niter, gypsum, and lime; and identification of some salts by their crystal-form. In the analysis of mineral waters and the solid residues obtained therefrom by distillation it was common also to test with acids, like nitric, as well as to heat them to determine any burning qualities and to note possible gases evolved.

Mineral water analysis had progressed extensively by the end of the seventeenth century. There were the tests mentioned above and more. Although there were numerous classifications of the waters in terms of properties, there existed almost no generally accepted scheme to analyze these waters. Prior to Robert Boyle there had been the tendency to apply these qualitative and quantitative tests almost solely to medicinal waters. Boyle extended the available techniques and, adding some of his own invention, applied them to all solutions, including sea water.

CHAPTER 2

ROBERT BOYLE

The *Observations and Experiments on the Saltness of the Sea* (35, vol.3, p.764) is Robert Boyle's only separate treatise that dealt solely with the sea and its water, although throughout his writings there are references to the sea and "sea salt".[28] The *Observations and Experiments on the Saltness of the Sea* contains a wealth of information concerning various aspects of the sea gathered from both experiments and interviews with mariners.[29] It is here that Boyle attempted to describe the saltness of the world's oceans both at the surface and at depth:

> The cause of the saltness of the sea appears, by Aristotle's writings, to have busied the curiosity of naturalists before his time; since which, his authority, perhaps, much more than his reasons, did, for divers ages, make the schools, and the generality of naturalists, of his opinion, till towards the end of the last century, and the beginning of ours, some learned men took the boldness to question the common opinion; since when the controversy has been kept on foot, and, for aught I know, will be so, as long as it is argued on both sides but by dialectical arguments, which may be probable on both sides, but are not convincing on either. Wherefore, saltness of the sea, obtained by my own trials, where I was able; and where I was not, by the best relations I could procure, especially from navigators.
>
> First then, whereas the Peripateticks do, after their master Aristotle, derive the saltness of the sea from the adustion of the water by the sunbeams, it has not been found, that I know of, that where no salt, or saline body, has been dissolved in, or extracted by water exposed to the sun or other heat, there has been any such saltness produced in it, as to justify the Aristotelian opinion.

The issue Boyle took with Aristotelian beliefs is obvious here although Aristotle's opinion concerning the cause of the sea's saltness is misquoted:[30]

> This may be gathered, as to the operation of the sun, from the many lakes and ponds of fresh water to be met with, even in hot countries, where they lie exposed to the action of the sun. And as for other heats, having out of curiosity distilled off common water in large glass bodies and heads, till all the liquor was abstracted, without finding, at the

> bottom, the two or three thousandth part, by my guess, of salt, among a little white earthy substance that usually remained. And though I had found a less inconsiderable quantity of salt, which, I doubt not, may be met with in some waters, I should not have been apt to conclude it to have been generated out of the water, by the action of the fire, because I have, by several trials purposely made, and elsewhere mentioned, found, that in many places (and I doubt not, but if I had farther tried, I should have found the same in more) common water, before ever it be exposed to the heat of the sun or other fire, has in it an easily discoverable saltness of the nature of common salt, or sea-salt, which two I am not here solicitous to distinguish, because of the affinity of their natures, and, that in most places, the salt eaten at table, is but sea-salt freed from its earthy and other heterogeneities, the absence of which makes it more white than sea-salt is wont to be with us. These last words I add, because credible navigators have informed me, that in some countries, sea-salt, without any preparation, coagulates very white; of which salt I have had (from divers parts) and used some parcels. (35, vol.3, pp.764–765)

By 1670 and before, a variety of authors (such as Paracelsus, with the number of elements) had taken issue with some aspects of Aristotelian doctrine (123, pp.168, 185). The *Observations and Experiments on the Saltness of the Sea* is notable here because it is the first work to openly take issue with Aristotle and not only with the description of the seas' saltness but also its cause. Boyle disagreed with Aristotle that the sun was the cause of the saltness and he attributed the salt remaining after evaporation simply to the salt that had been dissolved.[31] Boyle believed that the seas' saltness was due simply to the salt dissolved in the water and that this salt came from the earth and rocks, which river waters emptied into it. He knew that sea water was not nearly saturated with salt – having added on several occasions extra salt so as to demonstrate this fact (35, vol.3, p.776). Since the ocean was not saturated he saw no reason why saltness could not vary from place to place (35, vol.3, p.764).

Boyle strongly questioned the authority of Aristotle; but what is more important, especially to the emergence of science and chemistry, he said that the points that he delivered concerning the saltness of the sea were obtained by experimentation and by discourse with navigators, implying that Aristotle's were not.[32] Initially he based his speculations on numerous talks with divers who stated (by taste) that the sea was salty to the bottom; Boyle went on to test samples of sea water. Robert Hooke, the famous English physicist who was Boyle's assistant for a time, designed a two-valved copper sampling device (140) by which representative samples of the waters around England especially from the Channel were collected.[33]

A great number of Boyle's measurements of saltness were done by means of the hydrometer:

> It is uncertain who invented the hydrometer, it may possibly have been Archimedes.
>
> The first clear description of the instrument is given in the fifteenth letter of Syrenius (A.D. 370–415) to his friend and mentor Hypatia (appointed Bishop of Ptolemais, A.D. 410). "I am so unfortunate as to need a Hydroscope. Have a bronze one constructed and assembled for me. It consists of a cylindrical tube the size and shape of a flute. On this tube in a vertical line are the notches by which you tell the weight of the waters we are testing. A cone fits over the tube so evenly that cone and tube have a common base. . . Now whenever you place the tube in the water it will remain erect" (quoted in 258, p.7). Robert Boyle (1675) designed a glass hydrometer very similar in design to the modern instrument. (258, p.7)

These were done by himself but only on samples brought to him and by a number of mariners that he equipped with "hydroscope" glass instruments. Further tests were done using weighing techniques. He used phials with long straight necks[34] which he weighed and compared to one another. Pieces of sulfur were also weighed in surface and bottom waters. Although most of Boyle's measurements of saltness were physical he did employ two chemical methods. One was simply the evaporation of a pound of sea water (p.777, Avoirdupoise) in a digestive furnace. The second chemical method used was the addition of some agent supposedly to precipitate the salt. While Boyle, himself, classified the evaporation procedure as a chemical one; and although he got inconsitent values with respect to the salt contained in different pints of sea water (even from same area), he never suggested that this method should not be used. In fact, he felt that his inconsistent values (higher than expected) might well be the result of the moisture in the air. He suspected that the evaporation method was not as straightforward as one might expect, and furthermore he thought that some change in the salt might be occurring:

> From whence this greater proportion of salt by distillation, than our other trials invited us to expect, proceeded, seems not so easy to be determined; unless it be supposed (as I have sometimes suspected) that the operation the seawater was exposed to in distillation, made some kind of change in it, other and greater than before-hand one would have looked for; and that, though the grains of salt we gained out of the seawater, seemed to be dry before we weighed it, yet the saline corpuscles, upon their concreting into cubes, did so intercept between them many small particles of water, as not to suffer them to be driven away by a moderate warmth, and consequently such grains of salt may

> have upon this account been less pure and more ponderous than else they would have been. (35, vol.3, pp.777–778)[35]

It is not surprising that Boyle should have tried to measure a water's saltness by evaporation.[36] This had been the standard method from time-immemorial to procure salts. While seawater was evaporated to produce salt, certain natural waters were evaporated to acquire other salts, such as natrum (soda) from the waters of the Nile (1, p.559).[37] Since virtually all salts were separated from water by evaporation it is surprising that Boyle chose as his primary test (other than hydroscopic) a non-evaporated method.[38] Part of the reason for this is possibly the prejudice against chemical residues which appeared and was such a pronounced characteristic of the chemistry of the sixteenth century. It was strongly felt that the residue obtained after the drying process had undergone a change in the drying and was no longer the same composition as the material originally dissolved in the water (223, p.343).

There is only one actual chemical test mentioned and used by Boyle in the *Observations and Experiments on the Saltness of the Sea*. This precipitating agent was oil of tartar *per deliquium*[39] that he recommended:

> Having then, upon some of the distilled liquor, to be made; whereas a small proportion of that liquor being dropped into the undistilled sea-water itself, it would presently trouble and make it opacous, and though but slowly, strike down a considerable deal of a whitish substance (which, of what nature it is, I need not here declare). . .
> (35, vol.3, p.772)

He did say that sal ammoniac (ammonium chloride) worked, but not as well.

It is curious that Boyle should not have included the silver nitrate precipitation in the *Observations and Experiments on the Saltness of the Sea* as it was definitely the most sensitive test for salt in solution he had access to. In the *Philosophical Transactions* of 1666 he mentioned in a review of "Origins of Forms and Qualities" (p.195) the formation of "Luno corneo" (silver chloride) by spirit of salt (hydrogen chloride: hydrochloric acid) from a solution of silver. In his *Experimental History of Colours* (35, vol.1, p.757) Boyle described the process of dissolving silver in "aqua fortis" (HNO_3), concentrating it to remove the crystals so as to dissolve these in water to form a clear solution which would give a "very white precipitate" with an alkali[40] or an acid spirit.[41] At this point Boyle seemed to note no relationship between salt and spirit of salt with respect to their action on a solution of silver in "aqua fortis".

Silver calx, known under the name of *Luno corneo* (silver chloride) was known well before Boyle's time. Della Porta mentioned in his most famous works *Magia Naturalis* (first edition in 1561) that it may be made by grinding silver amalgam with common salt and distilling the mercury off, or by dissolving silver in aqua fortis and precipitation with a water-salt solution (240, vol.2, p.18; 255, p.174). Oswald Croll (1580–1609), a German physician, in his principal work *Basilica Chymica* (1609) (240, vol.2, p.175) described the precipitation and fusing of "luna cornea" (silver chloride). Van Helmont (1579–1644) later described the blue milky solution of precipitated silver chloride (240, vol.2, p.226).[42] Andreas Libavius (1540?–1616) gave a graphic description, in 1606, of the precipitate of silver chloride from a silver nitrate solution by that of salt solution (240, vol.2, p.255). The great German chemist and authority on salts Johann Rudolph Glauber (1604–1670) by 1667 in his *Pharmacopoea Spagirica* described the white precipitate of a silver nitrate solution by common salt (240, vol.2, p.356).

The precipitation of silver from its solution ($AgNO_3$) by salt was known in Agricola's time (1, p.443, note 10). Geber, whose works were printed in Latin by 1470–1480 (1, p.609), and the *Probierbüchlein* both describe it.[43] Agricola did mention certain purifications of silver (1, pp.439,443) that involved the precipitation of silver with salt. Hadrian Mynsicht (1603?–1638), a follower of Paracelsus, described it as "catharticum argentum" (240, vol.2, p.179) in 1651 as did Angelus Sala (1576?–1637) shortly prior to Mynsicht (240, vol.2, p.181). By 1650 the use of aqua fortis and the subsequent precipitation by spirit of salt (HCl) to form the white precipitate had been regarded for at least 150 years as an excellent test for silver as well as a means to purify aqua fortis itself.

Thus by the time Boyle had published the *Observations and Experiments on the Saltness of the Sea* not only was silver nitrate and its preparation well known, but silver chloride and its preparation[44] were equally well known. Since Boyle knew by 1666 that the "luna cornea" could be precipitated by spirit of salt, it is surprising that precipitation of this substance by silver nitrate in a salt solution (or in seawater) came so much later, especially since salt and spirit of salt were known to be related.

In 1683 Boyle wrote a letter to Dr. John Beal, a Fellow of the Royal Society (25, vol.4, pp.593–595).[45] Boyle mentioned that:

> For having long since written a short discourse of the saltness of the sea, I had been industrious to devise ways of comparing water in point of brackishness: and by these I found the patentees water to be more free from common salt, than waters, that are usually drank here in London.[46] (35, vol.4, p.594)

This indicates that between 1674 and 1683 he developed a new test for saltness and, he wrote:

> I found what possibly you will think strange, that if there were in water

> so much as one grain of salt in above two ounces of water, I could readily discover it; and the other, that even by this critical examen, I could not detect so much as a thousandth part of salt in our prepared water; whereas I found, by trials purposely and carefully made, that our English sea water contained a 44th or 45th part of good dry salt, of which is all one, that 44 pints, or near so many pounds of marine water, which would yield about one pound of dry common salt.[47]
>
> (35, vol.4, p.594)

From the accuracy as is given here the test had to be silver nitrate. By 1683 and later Boyle had a test for natural waters. Did he then see Tachenius' writings in the meanwhile?[48] Although Tachenius was fairly widely read on the continent it seems evident that Boyle had not seen this when he published his *Observations and Experiments on the Saltness of the Sea* in 1674. The letter of 1683 to Beal implies that Boyle arrived at this test himself. Such was most likely the case. There seems to have been sort of a step-wise development of his ideas. Some time earlier (by 1666) he knew that spirit of salt precipitated "luna cornea" as a white precipitate. This served either as a test for silver or spirit of salt (23, p.129). Later (after 1683) he simply seems to have reversed the spirit-of-salt test, enlarged this test, and considered the white precipitate formed with silver solution (silver in "aqua fortis") to be a test for salt.

Although his *Experiments and Considerations Touching Colors* of 1664 (35, vol.I) had dealt appreciably with color indicators and their use in some solution analyses, the *Short Memoirs for the Natural Experimental History of Mineral Waters* (35, vol.4, p.794), published in 1684–1685, represents most of Boyle's important as well as intervening and later ideas on mineral waters (75, p.60), and also represents the most comprehensive and complete treatise on solution analysis published to that time. Though Boyle had published the *Observations and Experiments on the Saltness of the Sea* in 1674 and the *Letters to Dr. John Beal* in 1683 (see p.21), they contained little on mineral water analysis.

Boyle stated clearly that the reason he again turned to the study of these waters was for "the health of thousands" (35, vol.4, p.795).[49] In Section I of these memoirs Boyle mentioned that (27, vol.4, p.795) work thus far on mineral waters had been narrowly confined to the oak gall tests[50] as well as some slight improvements in the technology of evaporation.[51]

The *Short Memoirs for the Natural Experimental History of Mineral Waters* is unique in that Boyle, in Section II, suggested the compilation of natural history for each specific water by investigation, and he gave a detailed 17-point procedure to aid in accomplishing this end (see Appendix I). The last section, number 17, is the only one of the group that is chemical in nature. It serves as sort of an introduction to section III which is the most important part, or the "physicochemical" part as Boyle calls it, of the "Short Memoirs".

> 17. Whether any thing considerable can be certainly discovered, or any very probable conjecture made, of the nature and qualities of the substances, that impregnate the water, by chymically and mechanically examining the mineral earths, through which it flows, or in which it stagnates? And particularly, by observing their colour, whether native or acquired, by being kept in the fire; their specific gravity; their affording or not affording, any salt, or other soluble substance, by decoction; their being soluble, or indissoluble, in particular chymical menstruums of several sorts, as *aqua fortis*, spirit of salt, etc. and their being committed to distillation in vessels of differing sorts, and various degrees of fire, with care to receive separately the differing substances they afford, whether in the form of liquors, or of flowers; and by examining these substances by fit and proper ways, as also the *cap. mort* by calcination, elixiviation, and (if it will bear such a fire) vitrification?

In part III Boyle gave the most complete scheme yet devised for the analysis of natural waters. He listed a total of 30 tests. The first 15 are as follows:

> 1. Of the actual coldness or heat of the mineral water proposed.
> 2. Of the specific gravity of the mineral water proposed.
> 3. Of the transparency, the muddiness, or the opacity of the mineral water.
> 4. Whether the mineral water will, by standing for a competent time, let fall of itself any oker, or other earthy substance, especially though the liquor be kept from the air?
> 5. Whether any thing, and if any thing, what can be discovered in the mineral water by the help of the best microscopes adapted to view liquors?
> 6. Of the colour or colourness of the mineral water.
> 7. Of the odour of the mineral water, as acetous, winy, sulphureous, bituminous, etc.
> 8. Of the taste of the mineral water, as acid, ferruginous, vitriolate, lixivial, sulphureous, etc.
> 9. Whether any change will be produced in the transparency, colour, odour, or taste of the mineral water, by its being taken up at the spring-head or other receptacle; or removed to some distance, by its being kept stopped or unstopped for a greater or lesser space of time; and, by its being much warmed or refrigerated, and also, by naturally or artificially produced cold, turned into ice, and thawed again?
> 10. Of the thinness or viscosity of the mineral water.

11. Whether the mineral water be more easy to be heated and cooled, and to be dilated and condensed, than common water?
12. Whether the mineral water will of itself putrify, and, if it will, whether sooner or later, than common water, and with what kind or degree of stink and other phaenomena?
13. Of the change of colours producible in the mineral water by astringent drugs, as galls, pomgranate-peels, balaustium, red roses, myrobolans, oaken leaves, etc. as also by some liquors or juices of the body.
14. Whether any thing will be precipitated out of the mineral waters by salts or saline liquors, whether they be acid, as spirit of salt, of nitre, *aqua fortis*, etc. or volatile alcalies, as strong spirit of urine, sal ammoniac, etc. or lixiviate salts, as oil of tartar *per deliquium*, fixt nitre, etc.?
15. How to examine with evaporation, whether the mineral water contain common salt, and, if it does, whether it contains but little or much?

While number 13 illustrates his continued interest in color indicators, 14 and 15 are for this study the most important. In 14 he mentioned the study of the water by precipitation and gave a number of specific precipitants (precipitating agents). In 15 there was the examination of waters for the presence of common salt. In neither of these did he mention the use of silver nitrate as a test agent. While he hinted at a test for salt in latter notes (35, vol.4, pp.813–814), in this 15th title, while mentioning his interest in the saltness of water (and specifically the King's command), he never gave any test for salt, other than the ability to identify it by its cubical grains (27, vol.4, p.816).

It seems reasonable to suppose that if Boyle had established the silver nitrate test for salt on what he considered to be a firm footing he would have mentioned it here, especially under title 15, Section III:

16. How to examine, without evaporation, whether the mineral water have any acidity, though it be but very little?
17. Of the liquor or liquors afforded by the mineral water by distillation *in balnco* and other ways.
18. Of the residence, *cap. mort.* of the mineral water, when the liquor is totally evaporated or distilled off; and whether the *cap. mort.* be the same in quantity and quality, if produced by either of those ways?

In number 18 he specifically mentioned the evaporation and distillation processes and questioned whether the residue produced by each is the same. While Boyle almost invariably specified distillation, this is the first time Boyle mentioned a possibility of different substances being produced by this process. This seems to be further indication of his belief that the residue underwent some change in being rendered such (see p.19).

The manufacture of artificial mineral water as suggested by Boyle in number 30 (below) was not a new idea. It is significant that he suggested it here since it is not only in keeping with his earlier attempt to make seawater by adding salt to water, but it is important in that future solution analyses did commonly attempt to do this, quite possibly due to Boyle's example.

19. Whether the proposed water being, in glass vessels exactly luted together, slowly and warily abstracted to a thickish substance; this being reconjoined to the distilled liquor, the mineral water will be redintegrated, and have again the same texture and qualities it had at first?
20. Whether a glass full of mineral water being hermetically sealed, and boiled in common water, deep enough to keep it always covered, will have its texture so altered, as to suffer an observable change in any of its manifest qualities? And if it do, in what qualities, and to what degree of alteration?
21. Of the proportion of the dry *cap. mort.* to the mineral water, that affords it.
22. Of the division of the *cap. mort.* into saline and terrestrial, and other parts not dissoluble in water, in case it contain both or more sorts.
23. Of the proportion of the saline part of the *cap. mort.* to the terrestrial.
24. Of the fixity or volatility of the saline part in strong fires.
25. Whether the saline part will shoot into crystals or not? And, if it will, what figure the grains will be of? And, if it will not, whether being combined with a salt, that will (as purified sea salt-petre, etc.) it will then crystallize; and if it do, into what figures it will shoot, especially if any of them be reducible to those of any species of salt known to us?
26. To examine, whether the saline part be, *ex praedominio*, acid, alcalisate, or adiaphorous?
27. Of the observables in the terrestrial portion of the *cap. mort.* as, besides its quantity in reference to the saline, its colour, odour, volatility or fixity in strong fire; its being soluble, or not dissoluble by divers menstruums, as spirit of vinegar, spirit of urine, oil of tartar, etc.
28. Whether, and (if anything) how much the mineral water's earth looses by strong and lasting ignition? What changes of colour, etc. it thereby receives? Whether it be capable of vitrification per se? And what colour, (if any,) it will impart to fine and well powdered Venice glass, if they be exactly mixed and fluxed into a transparent glass?
29. Of the oeconomical and mechanical uses of the mineral water, as in brewing, baking, washing of linnen, tanning of leather, or drying of

cloth, callicoes, silks, etc. as these may assist in discovering the ingredients and qualities of the liquor proposed.
30. Of the imitation of natural medicinal waters, by chymical and other artificial ways, as that may help the physicians to guess at the quality and quantity of the ingredients, that impregnate the natural water proposed.

The *Philosophical Transactions of the Royal Society of London* of 1693 contain a posthumous article by Boyle which was an extension of his letter of 1683 (October 30) to Dr. Beal (35, vol.4, pp.593–595). Here, for the first time, Boyle clearly described the silver nitrate test for salt made in a very weak solution of salt and distilled water:

> I took some common Water distill'd in Glass Vessels, that it might leave its Corporeal Salt, if it had any, behind it, and put into a Thousand Grains of it, one Grain of dry common Salt: Into a convenient quantity, for Example, two or three Spoonfuls, of this thus impregnated Liquor, I let fall a fit proportion, for instance Four or Five drops, of a *very strong* and well filtrated Solution of well-refined Silver, dissolv'd in clean *Aqua Fortis*; (for a shift, common or Sterling Silver will serve the turn:) And I made the Experiment succeed with Spirit of Nitre, instead of *Aqua Fortis*, upon which there immediately appear'd a whitish Cloud, which tho' but slowly, descended to the bottom, and settled there in a white Precipitate.[52] (33, pp.628–629)

Boyle went on to mention that this silver solution would be a useful experiment to test the sweetness of all waters, those of springs, lakes, rivers, wells, as well as seawater, especially where these waters might have possible use in the manufacture of beer, ale and mead. He attempted to quantify this test by mentioning that one could measure the amount of precipitate that four or five drops of silver solution would produce in a fixed amount of water having chosen a standard.[53] This silver solution according to Boyle was capable of detecting a grain of dry salt dissolved in 3,000 times its weight of water (25, p.628).[54] The development of this test was not considered by Boyle to have been very difficult:

> My way of examining the Freshness and Saltness of Waters, tho' (because it is wont to be surprizing the first time one sees it try'd, and has had the luck to be much talk'd of in many good Companies) 'tis thought to be an Invention very difficult, to be either found out or practis'd, is *yet* really no such mysterious thing as Men imagine it. And for my part, I hope it will be found much more considerable for its use, Than I think it is for the degree of Skill and Sagacity, that was necessary to devise it. For when I remembred and consider'd, that (as I have

found by various Trials) divers Metalline, and other Mineral Solutions could be readily precipitated, not only by the Spirit of Salt, but by crude Salt, whether dry or dissolv'd in Water, 'twas no very difficult matter for me to think that by a heedful application of the Precipitating Quality of common Salt, one might discover whether any Particles of it, (at least in a number any way considerable) lay conceal'd in a distill'd Water, or any other propos'd to be examin'd. (25, p.628)

It would appear from this that Boyle, in working with different solutions and in search of various precipitating agents both in themselves and as possible testing for purity of waters, arrived at the use of silver nitrate as a test for the saltness of water between 1683–1691. Although by 1666 he knew that spirit of salt (HCl) formed a precipitate with the silver solution and that spirit of salt was formed by acid action on salt, it took him several years until he tried salt solution itself with the silver solution as a test for a water's saltness.

By virtue of the fact that this test was hinted at in the Letter of 1683 to Beal, omitted essentially in the "Short Memoirs" of 1684–1685 and mentioned in detail in the addendum to the letter (25) of 1693, it can only be assumed that although this reaction was known some time prior to this, Boyle was not certain as to the validity of this test and used the intervening years to establish this validity.

Approximately 20 years after the *Observations and Experiments on the Saltness of the Sea* was published Boyle did present the silver nitrate test. While he *did not* single out any specific kind of water as being especially adaptable for this test, he was the first to offer clearly this as a test for purity of natural waters, and he was the first to apply its use to the analysis of seawater.[55]

Many of the tests that Boyle used were in use prior to his own time (23, p.126). Most of the early analyses, as Boyle himself indicated by his initial reliance on taste in the *Observations and Experiments on the Saltness of the Sea*, were based on the senses (as opposed to data arrived at by equipment or chemical means). As mentioned, the earliest true color indicator was the oak gall test. While a number of color tests such as color of salt solutions and some flame tests had been used by alchemists and metallurgists, the oak gall test was the only aqueous indicator which gave an actual color change in the presence of a specific substance (74, p.29). This simple test was to remain one of the important tests until well into the nineteenth century and was by Boyle's time the only indicator in general use for solution analysis. The first definitive established treatise dealing with color indicators (195, p.229) was written by Boyle in 1663: *The Experimental History of Colours* (35, vol.1, p.668).[56]

It seems clear that Boyle was interested in water analyses through his contact with his friends at Oxford. An acquaintance of Boyle's, Dr. Thomas Willis, in 1660 the Sedleian professor of natural philosophy and a well-known physician at Oxford, published a treatise on color indicators in 1659. These topics were known to and seemed to have been of interest to the Oxford group (75, pp.58–59). Why Boyle went to the study of the sea after 1664 ("Experimental History of Colours") and as indicated by the *Observations and Experiments on the Saltness of the Sea* and well before 1684 (the "Short Memoirs") is difficult to say.

A possible answer may have been Boyle's interest in salts. Much of Boyle's early work in chemistry dealt with salts. Throughout his work he was undecided as to the nature of a salt (23, p.150). He especially never had any clear idea that there might be some relation between a salt and its acid form in solution (23, p.84). Salts he considered responsible for the form of saltness. A precipitate from solution he thought to be the result of the presence of a salt. He recognized the existence of salts as being acid, alkali or neutral (23, p.151). Since it was known that mineral waters and especially the waters of the sea contained a good deal of salts it is probable that Boyle began his early study of mineral waters as an outcome of his interest in salts and later he progressed from this to the study of sea water. His latter discourse on the sea and the letter to Beal in 1683 was, of course, prompted by the King's request.

The sea according to Boyle was not saturated with salt. It was salty from top to bottom and there was no difference, in general, in the surface and bottom saltness from place to place. There could be some minor local variances as he believed, for example, in the existence of undersea springs and he knew of the effects of river run-off. The saltness of the sea was greatest in warmer regions like the tropics and least in colder regions, as approaching the poles, and also in areas of river run-off. Presumably this meant that the waters in the tropics were equal respectively in salt content from surface to bottom.[57] Boyle did take water at depth but his samples were probably never from deeper than a few meters.[58] Most of his data on this topic came from discourse with divers.

While it may seem obvious that the sea is salty from surface to bottom this was not necessarily thought to be the case in the seventeenth century. Boyle's writings on this subject, especially in the *Observations and Experiments on the Saltness of the Sea*, represented a step forward – and it was against Aristotelian doctrine. As mentioned previously the history of the analysis of mineral waters is important because it was the analysis of mineral waters that gave rise to solution analysis and the basic chemical procedures used to analyze seawater.

There is little question that Boyle wrote the first chemical treatise on the subject of the sea. The works of Robert Boyle as indicated especially in the

Observations and Experiments on the Saltness of the Sea do represent the clearest, most complete ideas with respect to the sea and its saltness written for some time to come.

CHAPTER 3

THE BEGINNINGS OF THE SYSTEMATIC STUDY OF THE SEA: HALLEY AND MARSILLI

There was a flurry of activity on the subject of mineral water analysis that began about 1666. This was centered primarily in England and to a lesser extent in France. A number of such treatises appeared in the *Philosophical Transactions of the Royal Society of London* and the *Journal de Savans.* From the very beginning of the *Philosophical Transactions* (1665) and the *Journal de Savans* (1665), and especially in journals that began later, there appeared accounts of fresh water analyses (mineral, spring, well and river) in almost every issue until well into the third quarter of the nineteenth century. Few of these, however, contained any sea water analyses. Some analyses of sea water did appear periodically but seldom in any number. Often a chemist would attempt to analyze sea water, but usually only as a special case of fresh water.

The French chemists of Boyle's time were not overly concerned with the analysis of sea water, but there was, as one might expect, the usual and considerable interest in mineral waters and their contents. French literature of the seventeenth century contained accounts of numerous analyses by those members of the Royal Academy interested in chemistry.[59]

Of this period the most outstanding French analyst of waters was Samuel Cottereau Duclos (d. 1715). Duclos was the King's physician as well as one of the initial members of the Academy (240, vol.3, p.12). Throughout his life he was interested to a large extent in solution chemistry. He carried out extensive investigations of French mineral and drinking waters investigating over 60 samples of water from sources (240, vol.3, p.11). In these investigations he included several analyses of sea water (81,84). Duclos was familiar with the fact that mineral waters contained a possible variety of salts. The work of Duclos is of importance since in the analyses of several mineral waters he said that he had identified not only common salt but also a nitrous and a sulphurous-like material (80, p.25; 82, p.123). He believed that many natural waters contained nitre (80, p.30).[60] Since mineral waters flowed over the earth's surface, he believed that they should contain all substances although he identified only a few. As a matter of course in his analyses Duclos rendered the solution dry and examined the residue. There was little attempt to weigh this residue, nor did he make any attempt to use precipitation methods to separate a substance as a residue.

In the separation of a residue Duclos used an almost closed retort (84, pp.388–389). Here he mentioned that an alkali might be used to precipitate the common salt and in so doing purify the sea water. This effect was due to the affinity that alkali seemed to have for salt (84, p.389).[61] He was familiar with the tests that existed up to that time and mentioned, for example, the oil of tartar test with mineral water (80, p.25). Duclos was also familiar with the works of Boyle and was responsible for the reviewing of his writings for the Academy (240, vol.2, p.497).[62] He became interested in the analysis of sea water due to accounts communicated to the Academy of several previous attempts to desalinate sea water. He knew that sea water weighed more than an equal volume of mineral water or distilled water (83, pp.321–322).

In all, Duclos represented a variety of new points with respect to mineral and sea water analysis. He attributed the bitter taste in salt water to the presence of a salt other than salt itself. Duclos did find a bitter salt[63] in mineral water and later in sea water (81, p.50; 84, p.387), but he did not determine its identity. Prior to this, a non-salty earthy or bituminous-like substance was used to explain this bitterness (210, p.36).

There were several contemporaries of Boyle who wrote concerning the sea. In 1684 a Dr. Martin Lister (1638?–1712) (182) stated clearly that there was a real difference between sea water and natural brine, as the salts yielded were not altogether the same. This statement was based on freezing experiments with natural and sea water. Francis Hauksbee (d. 1713?) wrote an article in 1709 on the densities of various waters including sea water and the effect of temperature on them (131).

HALLEY

The greatest contemporary of Boyle with respect to the subject of the sea was a man generally not so recognized. This was the great English natural philosopher Edmund Halley (1656–1742), remembered primarily for the comet which bears his name. Between the years 1687 and 1715 Halley published four articles in the *Philosophical Transactions* dealing with the sea. These were outstanding both in clarity of expression and in scientific achievement.

In the first of these papers, "An Estimate of the Quantity of Vapour Raised out of the Sea by the Warmth of the Sun" (124), Halley attempted to calculate, based upon pan-evaporation experiments, the amount of water rising from the Mediterranean.[64] He then calculated the amount of water that the Mediterranean received from its nine major rivers, each of which he roughly equated to the Thames which flow rate he had calculated from observation (124, p.369). The net effect was that his figures showed that more than three times the vapor left than was received per diem.[65] The validity of these values is not in question. What is important is the fact that here (and in later papers) Halley attempted to measure and calculate the water lost by the sea by evaporation and gained by subsequent river run-off.

Halley continued his investigation on this topic (125,126) primarily in an attempt to explain the difference in his previous (above) figures since it was evident to him that the sea had not changed its level in hundreds of years. The 1691 paper (125), "An Account of the Circulation of the Watry Vapours of the Sea and of the Cause of Springs," showed an excellent perception for the natural processes by which water rises from the sea primarily by the action of the sun and condenses particularly at higher elevations. Halley mentioned (125, p.471) that he noticed the condensation at higher elevations as hindrance to his celestial observations on St. Helena.[66] Halley clearly described the condensation of vapors to form springs which in turn form rivers and finally flow back to the sea.

A similar problem had been examined by the French savant Pierre Perrault (1608–1680) prior to Halley's calculations. For a three year period Perrault made rainfall measurements for the Burgundy region, where the Seine River rises, and decided that rainfall exceeded run-off by a factor of six (242, pp.449–450). Another French worker Edmé Mariotte (1620–1684), remembered primarily for his work on the chemistry of plants (240, vol.3, p.11), performed similar measurements for the region about Paris. He also investigated the penetration depth of rainfall and compared the water flow of springs in varying seasons (51, p.188; 317, vol.1, p.321).

The idea that the waters of the sea were recycled was hardly new. As such it had existed from antiquity (317, vol.2, p.361). The usual method given, however, for the rivers was by subterranean canals through which sea water passed and percolated and filtered through the earth until fresh whereupon emerging it became streams and rivers. This theory was prevalent at the time Halley wrote[67] and lasted to about the beginning of the eighteenth century. The work of Perrault, Mariotte and Halley was in the largest part responsible for its passing.

This recycling of the Earth's waters by evaporation and condensation was not a radical departure from nor contrary to existing knowledge and theory concerning the sea. By 1700 there was the general belief that the water of the Earth had always been there since the Creation (317, vol.2, p.351). It may have simply been put there in toto as such, or it may have condensed from the vapor state and accumulated in its natural place above the earth. Agricola, by 1556, had mentioned the rising of vapors of water (1, p.48). The origin and nature of the salt in sea water was not the subject of much controversy by 1700. It was generally believed that it had been dissolved out of the Earth's surface in regions of direct contact (317, vol.2, p.352). The salt in sea water was thought to be essentially common salt although it was well known that sea-salt had somewhat of a bitter taste and might not be as white in color as common salt. These apparently small differences attracted little attention and the accepted explanation was the presence of a bituminous material.

Halley's interest in the sea as evidenced by his papers was primarily that of a source of water vapor and the subsequent explanation of rivers and streams. He was, however, aware of the amount and saltness of the sea, and in a paper of 1715[68] (128) Halley proposed a method to determine the age of the world by virtue of the saltness of the sea and of several lakes. This original method was based on the observation that all waters contain at least some salt. According to Halley, lakes with rivers as their water source and which had no rivers issuing from them would become saltier, while those with rivers issuing in and out would tend to remain at about the same level of saltness. Rivers flowing into the lakes as well as rivers into the sea, explained Halley, carried salts. A comparison of the saltness of two lakes (one with a river emptying into it and another without) over a long period of time would provide by proportion the age of the earth. Halley believed it was possible that the saltness of the sea was increasing over long periods of time (2,000 years to be noticeable, 128, p.299) and felt that a monitoring of the saltness for a considerable time might also give a good indication of the age of the earth. Although not distinctly so phrased, Halley's idea of saltness and of the water itself is probably the first expression of residence times in the sea (128, p.299).[69]

While some might consider Robert Boyle to be the father of chemical oceanography (284), just as Edmund Halley might be classified as the father of physical oceanography, in the strict sense of the term neither of these men were oceanographers. Aside from the fact that the science did not exist as such then, the oceans were only of small interest to Boyle and Halley (and virtually all others previously) in the study of a variety of topics. Their papers concerning the sea occupied only a minor segment of their total works. Also, there is little evidence to indicate that Boyle ever went to sea. Others like Halley (and Aristotle) made only short trips over the sea.

MARSILLI

Count Louis Ferdinand Marsilli (Marsigli) (b. before 1690) was in most senses the first actual marine scientist.[70] Marsilli loved and was fascinated by the ocean, especially the Mediterranean, and he spent most of his life amid other occupations studying it. The largest part of Marsilli's life was spent under the aesis of France, although he was a general in Austria and a pensioner to Queen Christina of Sweden (51, pp.181–182). For some years he served as a military officer aboard ship in the Mediterranean. Throughout most of this time he spent his leisure in the study of the sea.[71]

The *Histoire Physique de la Mer* was written by Marsilli and published in Amsterdam in 1725. The great Flemish chemist Herman Boerhaave (1668–1738) bore the financial burden of its publication. The *Histoire Physique de la Mer* was Marsilli's last and most comprehensive but not his only writing on the sea. As early as 1680 he had sent his "Observations sur la Canal de Constantinople" (the Bosporus and the Dardenelles) to Queen Christina, and again in 1691 another communiqué on the subject. In 1711 Marsilli published in Italian a small 78-page book on the sea (209). This work was in reality the forerunner of the *Histoire Physique de la Mer.* Like the later version the "Brieve Ristretto Del Saggio Fiscio" was divided into five sections. There were passing comments on the color and temperature of the sea water throughout but virtually all of the coverage on the nature of the contents of sea water was contained in Part II. This comprised only four pages in the short 1711 edition.

Prior to the *Histoire Physique de la Mer*, no book existed that dealt solely with the sea from a scientific standpoint. It was the first book entirely oceanographic in nature. This book of 173 pages with many accompanying plates covered all of the aspects of the sea. The first part dealt with the sea's basin, the second with the water itself, the third the movements of the water, the fourth and the fifth marine plants and animals. Topics such as the nature of the bottom, saltness, temperature, density, currents, and color were treated. In the study of these he commonly made use of the hydrometer, microscope and balance. For density determinations at sea he used the hydrometer, as the balance was not reliable on board ship.

The second part of the 1725 *Histoire Physique de la Mer* summarized virtually all of Marsilli's ideas as to saltness and chemistry of the sea. In the 14-year interval between these two works Marsilli had added much material to this segment: 25 larger pages with many accompanying tables. More so than the 1711 edition, the edition of 1725 was primarily physically rather than biologically oriented.

The sea was divided by Marsilli into two regions: the surface and the deep (210, p.18).[72] Originally (in the 1711-edition) Marsilli had divided the depths of the sea into three parts: the surface, the middle and the deep. According to Marsilli (1711-ed.) the division was dictated by nature in that the surface waters were the lightest, and those of the bottom (deep) regions the heaviest with the waters of the middle regions lying somewhere in between. Since there was only a slight difference in the degree of saltness, Marsilli, in order to avoid confusion, later (1725) chose to divide the seas into only two regions, surface and deep.

The methods Marsilli chose to analyze sea water were essentially those of Boyle's. The chemical substances that Marsilli used on sea water were spirit of sal ammoniac and oil of tartar (210, p.25). In addition Marsilli used an "Eau de Fleurs de Mauvre" (210, p.25) as a standard test reagent.[73] With this solution Marsilli noticed the basic quality of sea water.[74] He was familiar with some sea waters that were acid (210, p.22), which he said was due to the presence of vitriol. Marsilli believed the surface waters of the sea to be more acid than those of the depths (210, p.31). He attributed this to nitre which existed in the air and which found its way into the surface waters rendering them more acid.

Rather than using a process of simple evaporation he, too, used distillation of known weights of sea water to produce the dry salt of the sea water.[75] This was not uncommon by this time (75, p.49). These residues he weighed. Over a period of time, in spite of all the care he took with the balance, he became convinced that the hydrometer was preferable in such measurements (210, p.25).[76] Aside from inconsistencies in residue weights Marsilli had a consistent lightness in weights determined by the balance. The weight loss Marsilli believed to be due to a loss of salt during the distillation (210, p.25) caused by the action of fire. The fire he believed actually consumed some salt. He found the same inconsistencies and lightness in using the balance to check saltness of artificially prepared sea water.[77]

Marsilli, in his attempt to study the surface and bottom regions, ran over a period of years analyses and specific gravity determinations for a number of water samples taken from these two regions for many locales. In so doing he noted that the Mediterranean water temperature was about the same from surface to bottom in the winter.

As early as 1680 the cause of the bitter taste in sea water had troubled Marsilli (210, p.22) and the primary aim of the second part of the *Histoire Physique de la Mer* (which treated of the water itself) was to explain this bitterness as well as to establish the degree of saltness (210, pp.19–20). Marsilli was aware that the degree of saltness varied from place to place, especially in the vicinity of rivers (210, p.18). Water was to Marsilli an element both tasteless and odorless and the salty and bitter tastes were purely accidental characteristics impressed on the water.

Surprising for a history of the sea, the *Histoire Physique de la Mer* contained a large number of fresh water analyses. One possible reason for this might have been Marsilli's attempt to isolate the bitter materials from a simpler water, namely fresh or mineral waters. It was a not uncommon feeling during Marsilli's time (and well after) that sea water was just a more concentrated and, therefore, more complex type of mineral water (195, p.61).

Initially Marsilli accepted a traditional view and explained the bitter taste of sea water as due to dissolved bituminous substance. After long and arduous work on this problem he added that the bitter taste came from the volatile bitumines which naturally could be driven of by the sun (210, p.36). He concluded that the surface waters were less bitter than those of the bottom. For Marsilli the nature of the bitumines that caused the bitter taste remained unidentified and he clearly stated this (210, pp.37–38). The very last comments he made on this topic were that the bitter material was injurious and cannot, as of yet, be separated from the sea water (210, p.41). He added that he thought possibly that the bitterness problem may have been solved in England but made no specific references. [78]

In the strict chemical sense it may be said that Marsilli contributed little to the chemistry of sea water. Yet perhaps this is a bit too harsh. Marsilli applied the use of colorimetric indicators to sea water and in so doing noted its basicity. He also pointed out the variances in the use of the balance in measuring the saltness of sea water samples as well as those differences as compared to the hydrometer.

The *Histoire Physique de la Mer* is truly a remarkable book. As with most creative thinkers, Marsilli's work is a complex mixture of the old and the new. While it was ahead of its time from the standpoint of ocean study its chemistry was characteristic of the period in which it was written. Marsilli believed, for example, in the Aristotelian notion that sea water could be rendered fresh by filtration (210, pp.32–33). The use of data tables was not new in water analyses although Marsilli first used them in reporting the results of sea water analyses. More significant was the reporting of a large number of sea water analyses along with the location from which each sample was taken, and accompanying tide, current temperature, and time data, much like modern station data. [79] Although on some points, such as the method he used to sample ocean depths, Marsilli was vague and generally omitted the description. He felt, however, that many of the ocean parameters influenced others and took great care in his measurement. Prior to Marsilli virtually all sea water samples were performed by people other than the sampler and brought usually some distance to the analyst.

The works of Count Marsilli did not contribute greatly to the concept of salinity, although Marsilli did emphasize the bitter quality of sea water and the concern that it should be identified. The *Histoire Physique de la Mer* was, however, a major contribution in that it presented a synthesis of the knowledge then available on the diverse aspects of the sea.

CHAPTER 4

FOUNDATIONS OF SYSTEMATIC MINERAL AND SEA WATER ANALYSES

The literature for some time after Marsilli contained very few articles on sea water at all. Those that did appear dealt primarily with attempts to purify sea water for drinking purposes.

In a treatise appearing in the *Philosophical Transactions* for the year 1753 there was an account of a process to render sea water fresh (306). The process itself is not of particular concern; it was simply a distillation method using calcined bones to hold the bituminous matter (i.e., the non-salt part) of the sea water in the heating flask (306, pp.69–70), thus maintaining the old belief that solutions needed some agent so as to be distilled. This paper did mention a number of tests used to check the purity of the distilled sea water. These were: the turbidity produced by sugar of lead (lead acetate), the turbidity with spirit of sal ammoniac (probably volatile ammonium chloride), the precipitate with tartar "per deliquium" (deliquesced potassium carbonate). Most important was the inclusion of the silver nitrate test:

> Into a spoonful of the distilled sea water he put twenty drops of a solution of silver in *aq. fortis*: He [a Mr. Appleby] likewise did the same with the like quantity of common water distilled. There appeared no change in either, and both retained their transparency.
>
> This demonstrates, that the distilled sea water is by the process entirely freed from marine salt, or its acid spirit. For, if we take a spoonful of common distilled water, and add the least particle of sea-salt, with the point of a penknife, and then drop into the mixture one or two drops of the solution of silver, it will appear turbid and milky.
>
> (306, p.70)

The sensitivity of this test for the sea-salt implied here is obvious. It is also important to note that it was recommended as the first test and thus presumably the best.

A later article, of 1765, also appearing in the *Philosophical Transactions*, dealt with, among other things, the determinations of the weights of fresh and salt waters (309). Contained therein were comments such as the fact that sea water was innately oily (309, p.99) and, because of this oiliness, was able to penetrate into substances like cork; that the sea became heavier as one proceeded away from land, especially from those shores with neighboring rivers (309, p.100). The most remarkable point was the inclusion of a comparison of the same vial (phial) weighed full of river water and then of sea water (309, p.103).

Although little was written on the subject of sea water during this period the literature veritably abounds with the analysis of mineral waters. As mineral water analysis gave rise to solution chemistry techniques and, therefore, the structure of later mineral and especially sea water analyses, its development is important in the story of knowledge of sea water even though it was some time until sea water analysis was considered somewhat apart from other water analyses, or until mineral water analyses had proceeded far enough so as to be useful in the analysis of sea water.

By and large mineral water analysis during the eighteenth century was a random thing. There was very little analytical pattern followed. The analyses did, however, become much more ambitious. Large volumes of single waters were evaporated, often as large as 16 "livres" (217, p.268),[80] and the residues were tested. Most of the famous mineral and spa waters were analyzed (298,267) and a number of substances in, and classifications for, these waters arose (139). By 1770 it was common to use acids in these analyses (217,139,298). The effervescence caused by the acid action was long since recognized (20, p.96). With such knowledge there was some basis of comparison of mineral waters.[81]

Virtually all of the analyses of natural waters to this time (1770) had evaporated the waters to dryness. A variety of tests, mostly with an assortment of tinctures, were then performed most often on the redissolved saline residue. However, a few new ideas appeared. The French savant Gilles François Boulduc (1675–1742) believed it more advantageous to separate the different substances contained in the water in the order they appeared during the evaporation process (24).[82] The great English natural philosopher Henry Cavendish (1721–1810) published in 1767 a treatise which was among the more notable of the time, "Experiments on Rathbone-Place Water" (52). It contained an estimate of the quantity of precipitated materials.

The chemists and physicians of these times were vigorously engaged in this task of mineral water analysis. Often it was repeatedly emphasized that these analyses were among the most difficult processes in chemistry. It says in the instruction published by the "Société Royale de Médecine":

> L'analyse des eaux minerales . . . est une des recherches chimiques qui exige le plus de ressources dans l'esprit de celui qui y applique.
>
> (195, p.71)

Many of these articles contained numerous analytical tests in lengthy analyses. They were, however, far from complete, and worse, they tended to be extremely inconsistent.

These inconsistencies are aptly pointed out by the great French chemist Antoine Lavoisier (1743–1794) in the opening sentences of one of his treatises on mineral waters:

> La partie de la Chimie qui porte le nom de Halotechnie, celle qui traite des Sels, est une qui semble avoir fixé l'attention des anciens Chimistes; l'analyse des Eaux minérales, qui appartient essentiellement à cette partie, s'est ressentie de ce retard; à peine y a-t-il cinquante ans que les Chimistes commencent à acquérir des idées nettes sur les différentes substances qui entrent dans leur composition, encore est-ce de nos jours que ces progrès ont été les plus rapides.
>
> Ceus qui se sont occupés particulièrement de cet objet, savant qu'il reste encore beaucoup à faire, et les différences énormes qui se trouvent dans les analyses d'une même eau, faites par différents Chimistes, prouvent combien cet Art peut encore prêter à l'arbitraire, ou au moins combien grande est l'extension des erreurs qu'on peut commettre: j'avoue que c'est quelquefois plutôt à l'Artiste qu'à l'Art qu'il faut imputer ce défaut de succès; mais il n'en est pas moins vrai qu'en simplifiant l'Art, on le mettra à portée d'un plus grand nombre d'Artistes.
>
> La difficulté de l'analyse des Eaux minérales, consiste principalement à séparer les différentes substances qui s'y rencontrent, à purifier les sels qui souvent sont imprégnés d'eau-mère, de matières extractives, ou de parties bitumineuses. (174, p.555)

Lavoisier added that although these analyses had interested the ancient chemists it was only in the previous 50 years that chemists had any clear ideas as to mineral water.[83]

By the year 1770 a complete analysis of a mineral water involved the examination of physical properties, qualitative reagent analysis, and analysis by evaporation (or distillation) which was becoming less popular.[84] This analysis was often subdivided into a treatment of the volatile materials and the residue. Lastly there was an examination of a mineral water's medicinal properties, usually based on case histories (283, p.204). It was not unusual for a section involving an artificial mineral water (prepared from substances found in the analysis) to be added to the total analysis. The treatment of the medicinal history of a water would often be omitted in a more chemically oriented analysis.

GIONETTI

Perhaps the best idea one can get in a brief way of the level of mineral water analysis by the late eighteenth century and a feeling for the increased sophistication in these analyses is to look at the work of the lesser known Italian chemist Victor Amé Gionetti (1729–1815). Gionetti, a medical doctor and a member of the Turin Academy of Sciences, published in 1779 a rather small book on mineral waters (112) which he declared contained some new analytical methods.

The residue of a distilled sample totaling 96 "livres" of water was found to be 4544 grains. This yielded a soluble mass of 3648 grains after extraction with distilled water. By evaporation and subsequent crystallization, Glauber's salt (sodium sulfate) was separated. Further evaporation, however, would not separate the other constituent salts which were, he thought, "sel commun" and "l'alcali de Soude". He then used two different methods to separate the material (112, pp.36–43). The solution freed of Glauber's salt was evaporated to dryness and weighed (139½ grains). Spirit of vinegar (acetic acid) of known strength (standardized against "l'alkali de Soude", pure sodium carbonate) was reacted with the mixture. Since he knew common salt (sodium chloride) did not react with vinegar, he thusly determined both.

> Or jugeant de la quantité de natron existante dans ces 139. grains et demi par la quantité d'esprit de vinaigre que j'avais employé, j'ai reconnu que toute cette masse était composée de 42. grains de sel commun et de 97. grains et demi de natron. (112, p.37)

Gionetti used a second method because he doubted the accuracy of the first:

> Je doutais d'ailleurs de n'avoir pas saisi au juste le point de saturation, quand j'avais employé l'esprit de vinaigre pour déterminer la quantité de l'alkali minéral, et enfin que l'impurité du sel de Soude, qui n'est pas certainement un sel alkali minéral éxempt de tout mélange de sels étrangers, pouvait n'avoir induit en erreur. (112, p.4)

This gave the same results as the first. Here he took 486 grains of the salt mixture and treated it with spirit of vinegar and extracted with spirit of wine (alcohol). The dried residue from the extraction containing the common salt and Glauber's salt weighed $357\frac{2}{3}$ grains. The alkali of soda (sodium carbonate) in the mixture was the difference, or $98\frac{1}{3}$ grains.

Although Gionetti used an acid of known concentration for the determination of the amount of "l'alkali de Soude" (sodium carbonate), his method was purely gravimetric rather than titrimetric (195, p.77), although in some of his methods he favored the titrimetric. For example, in one method he determined the sodium chloride (common salt) content of a residue by mixing it with alum, distilling off the hydrogen chloride gas ("gas acide marin") which was then dissolved in water to which sodium carbonate was added until basic.[85] The solution was evaporated to dryness and the weight of common salt determined (283, p.207).

Gionetti's work is important in the development of mineral water analysis in that he goes further than a number of his contemporaries, even farther than Lavoisier (195, p.77).[86] Furthermore, Gionetti checked the results of many different procedures with each other. The level of chemistry as evidenced in Gionetti's work is a good indication of that available at that time. As with most others of this time there was a conspicuous absence of an analytical scheme for the study of natural waters. Yet the increase in sophistication in water analysis as mirrored in and by Gionetti is evident from what it was just 30 years before.

LAVOISIER

Lavoisier himself seemed to be interested in the subject of water in virtually every way possible. Indeed it would be difficult to overemphasize the work of Lavoisier on water. Aside from his more theoretical and tremendously important works on the composition of water, he was interested in the composition of mineral waters, an interest which seems to have existed through most of his life. As a youth Lavoisier travelled extensively throughout France with his teacher, the prominent geologist Jean Etienne Geuttard (190, pp.46–52) who was preparing a geological map of France. Lavoisier found that the examined water generally contained a single salt: usually Glauber's salt (sodium sulfate) although often he identified "sel marin" (salt) or "sel gypseux" (selenite) (175, vol.3, p.163). Since he knew that the weight or density of water became greater with increased concentration of the salt, he reasoned that the difference between the density of the mineral water and that of distilled water would give a direct measurement of the amount of dissolved salt. Most of Lavoisier's measurements of salt content, then, were done with a hydrometer (areometer).[87] He took with himon these geological trips various reagents (whose concentrations he knew exactly), and weighed the reagent before and after so as to determine the amount used (195, pp.102–103). The concentrations of the mineral waters were determined hydrometrically relative to water. Later he relied on chemical methods and weighing techniques to a much greater extent. Much of this interest in things geographical presumably was due in large part to his friendship with Geuttard.

In 1772 Lavoisier wrote a paper on the use of alcohol in mineral water analyses (174). In it he chose to include the first analysis (Table I) of sea water ever published (174, p.563).[88] The reason he gave for this analysis was:

> L'eau de mer est le résultat du lavage de toute la surface du globe; ce sont en quelque facon les rincures du grand laboratoire de la Nature, on

doit donc s'attendre à trouver réunis dans cette eau, tous les sels qui peuvent se rencontrer dans le règne minéral, et c'est ce qui arrive en effet: comme cette eau est la plus compliquée de toutes celles que j'ai eu occasion d'examiner, je l'ai choisie pour donner un exemple de l'application de l'esprit-de-vin à l'analyse des eaux minérales.
(174, p.560)

Sea water, according to Lavoisier, was a mineral water, but the most complicated one that he had examined. For this reason he chose sea water as the example of the use of alcohol in mineral water analysis. The paragraph indicates Lavoisier's knowledge of the role of water in geochemistry.

The analysis of sea water was essentially this. Lavoisier evaporated the total volume of water slowly to dryness by means of a "feu de lampe" in a "capsule de verre" (174, p.560).[89] In the drying process "sélénite" and "sel gypseux" were precipitated naturally as the water became more concentrated. These salts were removed, dried and weighed. Alcohol was then added to the final dried saline mass and the "sel marin à base de sel d'Epsom" dissolved in it. The existing residue was then heated with a two-to-one mixture (by volume) of alcohol and water until completely dissolved. "Sel de Glauber" and "sel d'Epsom" crystallized from the cooled solution and were dried and weighed. The remaining alcohol-water solution contained some "sel marin" and "sel marin à base de sel d'Epsom" which was again slowly evaporated, dried, and weighed.[90]

Six years after the paper on the use of alcohol in water analyses Lavoisier wrote a short paper (194) on the analysis of water from the Dead Sea:

> Le lac Asphaltite est situé dans la Judée, sur les confins de l'Arabie pétrée; il est connu sous le nom de Mer morte; il est appelé dans la Bible, *Mer de sel, Mare salis, Mare salsissimum*. Cette dernière épithète annonce que les Anciens avoient reconnu que l'eau de ce lac étoit plus salée que celle de la mer. (194, p.69)

This he co-authored with Pierre Joseph Macquer (1718–1784) and Balthazar Georges Sage (1740–1824).[91]

The procedure for the analysis of the Dead Sea water (Table II) was essentially the same as that used in the sea water analysis previously (174, p.563). In the course of this analysis Lavoisier used eight different mixtures of alcohol and water. Lavoisier, Macquer and Sage found these waters to be high in "sel marin" and "sel marin à base de sel d'Epsom", but with virtually no "sels à base d'Epsom or à base terreuse ordinaire" (see Table I – only three or four grains).

Lavoisier described "sel marin à base de sel d'Epsom" in detail (194, p.70), noted its extreme deliquescence, mentioned its bitter taste, and the fact that with "l'acide vitriolique" (H_2SO_4) gave "l'esprit de sel" (HCl gas).

TABLE I

The results of the analysis of sea water by Lavoisier

On trouvera, en rapprochant les résultats rapportés ci-dessus, que l'eau-de-mer contient:

	pour 40 livres d'eau de mer			pour chaque livre d'eau de mer		probable composition
(1) Terre calcaire soluble dans les acides, et qui paroit ne pas differer de la terre calcaire commune		4	56		8-1/6	
(2) Sélénite ou sel gypseux						
	onces	gros	grains	gros	grains	
Sel marin à base d'alkali sixe de la soude	8	6-	32	1	54-4/5	sodium chloride
Sel de Glauber et sel d'Epsom	8	4	26		7-17/20	sodium and magnesium sulfate
Sel marin à base de sel d'Epsom	1	4	26	1	14-3/4	magnesium chloride
Sel marin, à base terreuse ordinaire, mêlé de sel marin à base de sel d'Epsom	1	5	10		23-11/23	calcium and magnesium chloride

(174, p.563)

TABLE II

The results of Lavoisier, Macquer and Sage for the analysis of water from the Dead Sea

En résumant ces expériences, on voit que l'eau du lac Asphaltite contient:

	par livre			par quintal	
(1) Sel marin ordinaire mêlé d'un peu de sel marin à base terreuse	1 once	0 gros	0 grains	6 liv.	4 once
(2) Sel marin à base terreuse composé d'environ quatre parties de sel marin à base de magnésie du sel d'Epsom, et de trois parties de sel marin à base terreuse ordinaire	6	0	57-3/5	38	2
	7	0	57-3/5	44	6

(194, p.71)[92]

He concluded that "l'acide marin" (HCl) was present in the composition of this salt. It is important to note that Lavoisier makes a point to say that the water of the Dead Sea contained not a single atom of bituminous substance. Other authors, he said, have wrongly attributed the bitter disagreeable taste of sea water to bituminous material when this property is due really to a variety of salts, but especially to "sel marine à base de sel d'Epsom" (magnesium chloride).

> Nous terminerons cette analyse en observant que l'eau du lac Asphaltite ne contient pas un atome de substance bitumineuse: c'est donc sans aucun fondement que quelques Auteurs ont attribué au bitume le goût mer et désagréable, soit de l'eau de la mer, soit de quelques eaux analogues; cette amertume est propre au sel marin à base calcaire et surtout à celui à base de magnésie ou de terre du sel d'Epsom (194, p.72).

Prior to Lavoisier nobody had explained the bitter taste of sea water in this manner.

Lavoisier is recognized as a major contributor to the chemistry of sea water (116,258), although he wrote only one article (194) solely on this subject and included only one other analysis of sea water in his writings (174). These two references do not represent his only work on sea water. Sprinkled throughout his collected writings are a variety of comments on this topic, although these usually were made in conjunction with the discussion of mineral waters. For example, in the *Traité Elémentaire de Chimie* (175, vol.1), published in 1789, he mentioned clearly that sea water contained very abundant "sel marin à base de sel d'Epsom" ("magnésie" combined with "l'acide muriatique") (175, vol.1, p.121).[93] In a long treatise about distillation he said that the distillation of sea water to produce pure water was possible (175, vol.4, pp.729–739).

Lavoisier was familiar with the precipitate formed in a salt solution by the addition of "dissolution d'argent" (silver nitrate) and he, of course, knew that the precipitate was "lune cornée" (luna cornea: silver chloride). In a treatise which dealt simply with the freezing of natural and artificial sea water "l'autre une pinte pareille d'eau de rivière, dans laquelle j'ai fait dissoudre deux onces de sel marin" (175, vol.5, p.240). He used this test as a rough indication of the saltness of the liquids (after some freezing had occurred). He stated that both waters seemed to be slightly less salty on the basis of this test and by taste (175, vol.5, p.240).

Evidently, Lavoisier did not regard this test as useful in that although he was quite familiar with it he only once used or mentioned it in reference to mineral water or sea water analyses. There was in regard to this test no mention of Boyle.

One paper of Lavoisier's is of particular interest here in that it contains his only really descriptive passages concerning the sea. These were contained almost casually in a paper dealing with the weights of different waters (175, vol.3, pp.456–460).[94] The sea water off of Cadiz was lighter than that of the high seas. This was due to the lighter water of rivers or streams mixing with the water at the coasts (175, vol.3, p.459). In a voyage extending a difference of 88° longitude and 8° of latitude, Chappe said that there was only slight variation in weight of sea water. Further, if one considered this slight relative variation, it seemed to indicate that the weight of the water diminished in going to the west from the east.[95] On this Lavoisier said that l'abbé Chappe had not proceeded far enough to conclude anything very positive as to the increase or decrease in saltness in approaching the equator. According to Chappe the saltness appeared to decrease, rather than increase in this direction, at least in the waters he had travelled.[96] Wihin sight of the shores of California the weight appeared to increase noticeably. This according to Lavoisier was a single observation which, when the occasion presented itself, should be repeated. Lavoisier explained the phenomena as due to the evaporation and rising of the waters in that region which was greater than that water received from the shores by rivers.

The results of l'abbé Chappe are at best indecisive. Lavoisier appears to have included them because he was interested in the subject matter and there was little else available in the literature on this topic. Shortly thereafter the Royal Academy of Science, probably as a result of l'abbé Chappe's work, but more so Lavoisier's comments, undertook the cost of supplying hydrometers similar (though not quite as accurate) to l'abbé Chappe's to travelers so inclined who were about to embark on travel that involved large longitude and/or latitude differences.

Lavoisier was not specifically interested in the sea any more than he was in a variety of other fields of possible endeavor. His interest in geology probably led him in some ways to the sea, but it was primarily water and natural waters that brought him to the study of sea water.

BERGMAN

There was one other contemporary of Lavoisier whose studies of natural waters were important. This was the great Swedish chemist Torbern Olaf Bergman (1735–1784). At the age of 32, Bergman was chosen to fill the chair in chemistry at the University of Upsala. This is surprising in that he possessed little of a formal chemical background even for those times. Nine years prior to this, however, he had been appointed as lecturer in physics, again with the same seeming lack of qualifications. This evidently was no limiting factor in physics as he quickly became well known in this field. Bergman must have been an exceptional person. Already recognized internationally as an outstanding zoologist, physicist, geographer, and astronomer (15, p.11) when he accepted the chemistry chair, he then proceeded with such a remarkable ability and grasp for this new subject that well before his death, 20 years later, he was recognized as one of the world's foremost chemists.

In his native Sweden Bergman is regarded as the father of the Swedish mineral water industry (15, p.111). This claim is based primarily on the large number of papers he wrote on mineral water in which he went to great detail both in analyses and subsequent syntheses. In this regard, of special importance are two articles, his "Bitter, Seltzer, Spa, and Pyrmont Waters and their Synthetical Preparation" (15) and "Of the Analysis of Waters" (14, vol.1, p.91).

It is not at all clear when Bergman began to show an interest in the chemistry of waters. The thesis "On the Waters of Upsal" (14, vol.1, pp.193–209) was the original documental evidence of this interest.[97] With the aid of Bergman's laboratory notes it would appear that he began to work with water analyses to some extent about May of 1769 (15, pp.111–112). Since Bergman had little to do with chemistry at all prior to 1767 this paper would be about the earliest possible beginning of interest.

There are a variety of possible reasons to explain Bergman's interest. Much of the impetus for his work seems to have been utilitarian. Since water was used in large quantities for purposes as well as for food, brewing, drinking, and cooling, he felt it necessary that more should be known about these waters. Analyses of natural waters would enable one:

> 1st, To chuse the purest water for internal use.
> 2nd, To avoid such as is either unfit or noxious.
> 3rd, To form a proper judgement concerning such as are useful in medicine. Thus, if long experience has shewn the efficacy of the water in a certain fountain, and if at the same time the contents of that water be known, we are enabled to anticipate the experience of years, and instantly to form a judgement concerning the virtue of other waters, which exactly resemble in their contents the water whose properties are already established.
> 4th, To select such waters as are best adapted to the several arts and manufactures.
> 5th, To amend the impure (in scarcity of good water) and to separate from it those heterogeneous particles which chiefly impede its use.
> 6th, To imitate such as are celebrated for extraordinary virtues, if a sufficient quantity of the natural water cannot conveniently be had.
> (15, p.108)

Furthermore, by the end of 1771, Bergman, on the advice of his doctor, began to regularly imbibe mineral waters. Since he thought them expensive, he wished to examine mineral water in order to make them artificially (15, p.115). Bergman also knew that considerable money left Sweden each year with the importation of mineral water from abroad. The analyses, synthesis, and subsequent manufacture of these waters would save the country a considerable amount each year.[98]

Bergman chose to study all natural water – that in air, snow, rain, springs, rivers, wells, lakes, marshes, and the sea. He made a point to look for bituminous oils and mentioned that nowhere in all these waters could he find any traces of such substance (14, p.10). Bergman made one of the first clear references to matter in water other than that which might be actually dissolved (i.e., suspended particulates). He was aware that very small particles of flint, lime, and clay were often found mechanically suspended in water (14, p.110). A comprehensive list of substances that might be found in natural waters in "chemical solution" (14, p.111) was also given. This was easily the longest list compiled by anyone by 1784 including Lavoisier.

According to Bergman water was capable of being analyzed by two methods: precipitation and evaporation.[99] These analyses of waters he considered to be one of the most difficult problems in chemistry. This was due to the presence of some materials in extremely small amounts (14, p.109). Added to this was the fact that the various dissolved salts, although often very dissimilar in nature were extremely difficult to separate from the mixed state in which they occur in mineral waters (14, p.171). Because of these problems, Bergman felt that the analyses should be confirmed by the synthesis of the mineral water based on the analyses (14, p.182).

Over several years Bergman developed a qualitative and quantitative scheme for analyzing natural waters. Initially the physical properties of the water were recorded and parameters such as clearness and temperature included. For temperature determinations an accurate thermometer was recommended and it was advised that the temperature measurements should be taken over the period of one year in the water's natural environment, in situ.

The water sample was then tested with a number of possible precipitating agents. This was to determine the contents and other properties such as acidity. Included among these precipitants[100] were the usual agents of the time, such as alum, nitrated mercury, corrosive sublimate, acetated lead, and nitrated silver,[101] as well as alcohol and soap.[102] These may be used directly on the new water sample or after it had been reduced in volume by evaporation (14, p.142). The study of waters by precipitating agents was only qualitative for Bergman.

The actual quantitative analysis was done by an evaporation method.

Generally one "kanna"[103] was *slowly* evaporated lest the rapid boiling cause some of the contained substance to be boiled off or even decomposed (14, p.156). The salts will separate out individually as the solution reduces in volume, the least soluble appearing first.

> If saturated solutions of different salts be mixed, they all appear, during the evaporation, in an order conformable to their degree of solubility; that is, such as are least soluble in water appear first. (14, p.158)

This is a really clear, simple statement of this fairly well known phenomenon. The entire residue, carefully weighed, was separated into insoluble part, the segment soluble in alcohol, and water soluble segments and separately analyzed.[104]

What is of primary concern here is Bergman's contribution to the analysis of sea water. Since he sought to analyze all natural waters, that of the sea was included. His results for the sea water analysis were given in "Dissertation of Sea Water" (14, vol.1, pp.226–231).[105]

The alkalinity of sea water was noted by Bergman. After the use of several indicators he determined that paper tinged with tincture of Brasil wood (14, p.227) was slightly altered. He stated that sea water was only weakly alkaline. This alkalinity he believed to be caused by dissolved magnesia.[106]

A *kanna* of the sea water was evaporated to dryness producing a total residue of 3 ounces, 378 grains. This residue was washed with alcohol and redried. The alcohol solution contained the salited magnesia (magnesium chloride).[107] The residue was extracted with cold water which dissolved and carried off the salt and the small remaining residue was gypsum. He tested for the presence of vitriolated magnesia (magnesium sulphate) in sea water and found it to be absent.[108]

Bergman gave the results as shown in Table III.

The water used by Bergman was collected in the region of the Canary Islands.[109] Well-corked narrow-necked glass bottles were used to collect and transport the samples. These were lowered, by means of weights, to a depth of 60 fathoms. Earlier trials of a depth of 80 fathoms had cracked one bottle.[110] The theory of this method is that as a corked bottle was lowered over the side, at some point (not necessarily the depth at which it was lowered) the cork would be forced into the empty bottle by the pressure. Once the bottle was raised the cork, floating in the water-filled bottle would re-enter the neck of the bottle thus trapping the contained water.[111]

Bergman knew that different mineral waters existed and that these not only contained different substances but also varying amounts. Yet his ideas on the nature of sea water show no indication that he felt this water might vary in composition. Bergman, like Lavoisier, regarded sea water as the most complex water in the physical world, and chose to analyze it in his tests of all forms of natural water.

With the exception of surface waters which he felt were tainted with matter from organisms, he apparently believed that the actual analysis of sea water samples gave identical results. Bergman did use a number of sea water samples, yet he published only one data table on sea water. Either he performed only one analysis of one sample, or mixed the samples and analyzed them. Since Bergman was, however, a superior chemist it is much more likely that he performed a number of analyses on a number of different water samples. From this it would appear that he determined no difference and then felt that any analysis of a sea water sample would give the same results. If the salt content was the same then the constituents would have to be constant. This is a necessary conclusion. Nowhere did Bergman explicitly state that the constituents that make up sea water were in constant proportion. But throughout his treatise on sea water and in some of his other works there is the suggestion that he believed the proportion to be constant.

It is curious that Bergman never seemed to detect any metals such as iron. Precise tests for iron existed, such as gallic acid tincture or blood-lye for example (ferrocyanide of potassium) (15, p.55). Either the high concentrations of salts masked the test or, more likely, he did not try these tests although he normally did them for mineral water.

Bergman's analysis of sea water represents simply the application of the scheme he devised. There was no attempt to single out sea water specifically for study. His method of analysis was a "wet" one.[112] He considerably expanded this method of assay and *introduced the procedure of weighing the precipitated salts*, a practice then not in common usage (279, p.137). In his methods he placed more emphasis than his contemporaries on tests and weighing. Not only did he realize the value of detecting and controlling the acidity of mineral water formulations *but he felt that the results of any water analysis might not necessarily indicate the actual constituents as they existed in solution*. Bergman never gave any reason for this feeling beyond the comment about the possibility of substances being changed or destroyed during evaporation, and thus the necessity of confirming the analysis of a water by synthesis (14, vol.1, p.182). There is no indication that he ever tried to do this with sea water.

Bergman is important in the history of chemical oceanography, not as an oceanographer but as a chemist. The inclusion of precipitation as a method of analysis is important. Largely due to Bergman's recommendation its usage continued to grow in the future. Even less so than Lavoisier there is nothing to indicate that he was particularly interested in the study of the sea even though Sweden was a maritime nation. His sea water analyses were done like any other natural water he might have chosen. Sea water was just another water.

Chemical analysis had existed for some time prior to Bergman, but it was

TABLE III

The results of the analysis of sea water by Bergman

Upon collecting and weighing all the contents, each "kanna" is found to contain:

	ounces	grains
of common salt	2	433
of salined magnesia	0	380
of gypsum	0	45
	3	378

(14, pp.229–230)

not a separate branch of the science. Bergman was responsible for its attainment as a separate branch of chemistry (analytical) (283, p.71). What is important here is the fact that Bergman wrote the first major scientific treatise on the analyses of natural waters and a major treatise of analytical chemistry. More importantly, contained therein was the first systematic analytical qualitative and quantitative scheme for solution analysis.[113] Bergman's work, however, was not a classification but truly a means to analyze and then classify if one so wished.

Indirectly, then, Bergman was important to chemical oceanography. His work on mineral waters stimulated and accelerated much of the future work on natural waters and solution chemistry.[114] It was the study of mineral waters that developed the techniques that enabled sea water to be analyzed.

CHAPTER 5

SEA WATER ANALYSIS AND THE PRECIPITATION METHOD

In only a short time after the publication of Bergman's works on water analysis (14) they became well known and were used by chemists almost immediately. Johann Christian Wiegleb (1732–1800) was inspired by Bergman and recommended strict adherence to his method of mineral water analysis (195, p.78). The outstanding French chemist Antoine de Fourcroy (1775–1809) was familiar with, and complied with, most of Bergman's recommendations (195, p.79). In England a number of prominent scientists made use of his method. The great Scottish chemist Joseph Black, M.D. (1728–1799) analyzed the water from a number of places including several hot springs in Iceland (20, 21). William Withering (1741–1799), a botanist remembered primarily for his classic work on foxglove (240, vol.3, p.300), also tried his hand at mineral water analysis (316). The method of analysis in these last two cases, with the exception of the addition of several qualitative tests, was almost identical to that of Bergman. Both Black and Withering, after an extensive series of qualitative tests, precipitative in nature, went on to evaporate the samples and separate the components primarily by solvent extraction.

Another Englishman, the noted chemist Richard Kirwan (1735–1812), well known for his outstanding work on the analysis of minerals between 1790 and 1810, became interested in mineral water analysis. Kirwan wrote a book in 1799 (157) in an attempt to render Bergman's system of water analysis simpler and quicker (283, p.114). The book was notable and valuable historically in that it contained a comprehensive list of references of all the work done in water analysis since Bergman. But there was little in the book that was really new. Possibly this was due to the restrictions dictated by the adherence to an existing scheme which was already rigid in its approach to water analysis, at least after the initial qualitative tests had been completed.

Kirwan gave a simple method for the determination of the total salt content of a water:

> There is a method of calculating the quantity of salt in 1000 parts of a saline solution whose specific gravity is known, which, however inaccurate, is yet useful in many cases as the error does not exceed 1 or 2 per cent, and sometimes is less than 1 per cent. It consists simply in sub-

tracting 1000 from the given specific gravity expressed in whole numbers, and multiplying the product by 1.4. It gives the weight of the salts in their most desiccated state and consequently freed from their water of crystallization. The weight of fixed air [carbon dioxide] must also be included, thus for a solution of common salt having its specific gravity 1.079, I find the difference from 1000 is . . . 79 and 79 × 1.4 = 110.6, then 100 gr of such solution contain 110.6 gr of common salt. . .

(157, p.145)

This was both new and different and came at a time when chemists had for some time only considered total salt content as that present after complete evaporation of the water. This method would be, in theory, applicable to any water including sea water.

By 1788 Bergman's influence in water analysis had spread to America (76). His analysis of sea water was often quoted (76;180, p.254), and sometimes cited as the definitive analysis (134).

Aside from the Bergmanian influence, water analyses all show by 1800 a marked advance in the use of qualitative tests. The actual weight determination of the components of a mineral water were still done by solvent extraction. These precipitating agents were only used to indicate the presence of certain specific substances. Among the most sensitive precipitating agents were oxalic acid, barium chloride (muriate of baryte) and silver nitrate, and these were all known and used by Bergman. Oxalic acid was his "acidum sacchari" and was the test for lime, barium chloride was "terra ponderosa salita" and was the reagent for vitriolic acid (H_2SO_4) and Glauber's salt (Na_2SO_4); silver nitrate was known to Bergman as "argentum nitrum" and was the test for (rock) salt.[115]

In following the development of mineral water analysis from Bergman to the early 1800's, the work of the renowned Scottish chemist Joseph Black (1728–1799) is a good bridge (20,21) for Black gives a good picture of the analytic chemistry of the time. In a variety of cases such as his use of slow heating in the evaporation of waters, he was aware that a rapid or high heat exposure would change the composition of the residue and that "part of the acid of the saline compounds" might be driven off (20, p.97). In the quantitative determination of constituents in a water sample Black used one very important departure from Bergman's method. This was the practice of drying to constant weight. Black did not initiate this method but he and other chemists of this time were quick to adopt it seeing easily its obvious advantages.

The practice of drying to constant weight was initiated by the German chemist Martin Heinrich Klaproth (1743–1817) (176, p.131), one of the outstanding analytical chemists of his time. Prior to the use of this practice it is difficult to imagine that any analysis even of similar water samples except very simple ones could begin to show some agreement. It is important also to note here that Klaproth introduced the practice of reporting the actual percentage composition based on analysis instead of recalculating the results in order to get a sum of 100% as was usually done. This method is not only more reasonable but it permitted the discovery of errors more readily (176, p.131).

The errors innate in the fractional crystallization and separations of salts in mineral waters were evident to Klaproth who devoted most of his life to analysis. He analyzed many minerals and simplified and refined techniques for their analysis, as well as discovering uranium, zirconium and cerium (283, p.119). His interest in mineral water evidently arose from his work in mineralogy. In improving the established method of water analysis he stressed the omission of fractional crystallization where possible:

> So I have devised a more reliable method which first of all involves the saturation of the free mineral alkalis with acid, and then I decompose the neutral which are formed with a suitable reagent; at the same time I carry out a trial experiment in order to make clear the relationship, and on the basis of this I calculate the result. (translated in 283, p.122)

He was not content with the evaporation and fractional crystallization techniques generally then used as the major separation and identification methods in water analyses. His new method was largely a precipitation procedure using mineral acids. Klaproth's analysis (about which more will be said later) included precipitation with sulfuric acid but solvent extraction was used to determine the proportion of magnesium and calcium chlorides (muriates) which he felt was the composition of the salt remaining after alcohol evaporation (158, p.38). It is surprising that Klaproth, in an analysis of the Dead Sea water in 1813, depended primarily upon previous evaporation methods (158). In addition, he accepted (158, p.39) Bergman's analysis of sea water which was done over 30 years earlier – especially surprising since Klaproth clearly attempted to play down fractional and evaporation techniques as being inaccurate, and Bergman did rely primarily on these methods. The method Klaproth used was largely similar to that Lavoisier had used some time earlier (194), although he refined it and implemented some precipitation techniques to a large extent. [116] These are primarily qualitative rather than quantitative. Klaproth's results were consistent within themselves due to his own common practice of drying to constant weight. There was no indication given as to why in the Dead Sea water analysis he chose almost to duplicate in procedure earlier attempts. The answer probably lies in Klaproth's own character and his regard for the work of Bergman.

The last part of the eighteenth century and well into the nineteenth produced no in-depth study of the sea although it was, and has, always been the subject of countless descriptive observations. Most of these descriptions had been made by navigators and were of a practical nature and usually were used in making voyages simpler. No purely oceanographic expeditions had been outfitted and most measurements taken were at best only scientific curiosities. The chemical investigations of the nature and constituents of sea water continued to be the only scientific study of the sea. There is no attempt here to imply that the chemical study of the sea was truly systematic at this time. There was no other study of the sea in this period. The analysis of sea water was at least systematic in that it usually followed a scheme. Even though this study of sea water was still considered to be a branch of mineral water analysis, and although it was still a recent endeavor, no other systematic study of the sea was taking place during this period. The only possible, though doubtful, exception might be the random collection of mineral and biological specimens deposited almost on the doorsteps of the newly blossoming museums throughout the world by so interested travelers abroad.

A small number of sea water analyses, however, were performed in the last and first quarters of the eighteenth and nineteenth centuries, respectively. The German chemist and botanist (240, vol.3, p.688) Heinrich Link had analyzed water from the Baltic Sea near Doberan and got the results as shown in Table IV. Shortly afterwards J.F. Pfaff also analyzed Baltic Sea water and published the results as summarized in Table V. These two results

TABLE IV

The results of the analysis of Baltic Sea water according to Heinrich Link

Salzsäure Bittererde	(magnesium chloride)	231,25 Gran
Schwefelsäure Bittererde	(magnesium sulfate)	4,166
Schwefelsäure Kalkerde	(calcium sulfate)	25
Kochsalz	(sodium chloride)	509
Harzige Substanz	(resin-like substance)	2
		771,416

(180, p.256)

TABLE V

The results of the analysis of Baltic Sea water by J.F. Pfaff

Kochsalz	(sodium chloride)	56 Gran
Salzsäure Kalkerde	(calcium chloride)	24
Salzsäure Bittererde	(magnesium chloride)	6
Gyps	(calcium sulfate)	6
Kohlensäure Kalkerde	(calcium carbonate)	1
		93

(249)

are quite different. The Baltic is a fairly dilute body of sea water and it is possible that in a region of river run-off both of these analyses could vary in the relative amounts.[117] Yet even the salts determined were not the same.

F.D. von Lichtenberg published another analysis of Baltic water in 1811 (180). The water was taken near the mouth of the Vistula river (which accounts for his low specific gravity of 1,006). Aside from containing a determination of carbon dioxide ("Kohlensäure") in sea water, this analysis is unusual in that it contained the important hint that the ingredients and relative amounts should be relatively consistent. "Man sollte aber meinen, die Verhältnissmengen der *einzelnen Bestandtheile* wurden sich, nahe wenigstens, gleich bleiben" (180, p.257). Lichtenberg did believe, however, that there might be some relation between salt content and atmospheric conditions such as wind direction:

> Und werden diese Veränderungen nicht nach bestimmten Gesetzen erfolgen; und diese Gesetze sich auffinden lassen? Es würden freilich dazu zahlreiche fortgesetzte Untersuchungen erfordert werden, die immer auf die gleiche Art angestellt werden, und wozu sich mehrere, in verschiedenen Gegenden an der See Wohnende, verbinden und zugleich auf die jedesmalige Beschaffenheit der Atmosphäre und der See, nach den verschiedenen Rücksichten, sehen müssten. (180, p.257)

His request for analyses of sea water by a host of workers from different areas under the same conditions and by the same manner was an important suggestion.

The German chemist H. August von Vogel, who worked extensively in France, and the French chemist E.J.B. Bouillon-Lagrange (1764–1844) coauthored an article on the sea waters that touched the shores of the French empire (29). Bouillon-Lagrange had previously analyzed a number of mineral waters of France and was familiar with water analysis as was Vogel (28).

The analysis of Bouillon-Lagrange and Vogel was of the waters of the English Channel, the Atlantic and the Mediterranean. They expressed the belief that although these bodies of water communicated with one another, there was a variance of content with latitude:

> Quoique les trois mers, dont nous avons fait l'analyse de l'eau, communiquent entr'elles, nous avons supposé une différence par rapport au degré de latitude. (29, p.507)

Also included here was the comment that there might be some difference in the quantity of salts contained rather than in the nature of the salts:

> . . . et nous étions tentes de croire que si ces eaux ne différaient pas par

TABLE VI

The results of analyses by Bouillon-Lagrange and Vogel for sea water from the English Channel, Atlantic Ocean and Mediterranean Sea

Nom des eaux	*Poids*	*Résultats de l'évaporation*	*Gaz acide carbonique*	*Muriate de soude*	*Muriate de magnésie*	*Sulfate de magnésie*	*Carbonate de chaux et de magnésie*	*Sulfate de chaux*
Eau de la Manche	1000 grammes	36^{g} de matière saline	$0^{g}.23$	$25^{g}.23$	$3^{g}.50$	$5^{g}.50$	$0^{g}.20$	$0^{g}.15$
Eau de la Mer atlantique	1000	38	0.23	25.10	3.50	5.78	0.20	0.15
Eau de la Méditerranée	1000	41	0.11	25.10	5.25	6.25	0.15	0.15

(29, p.515)

> la nature des sels, elles pouvaient varier du moins par la quantité des matières salines tenues en dissolution. (29, p.507)

Implied here (in 1813) almost offhandedly, was the *idea of constancy of specific salts in sea water*.

The results obtained by Bouillon-Lagrange and Vogel are summarized in Table VI. With a few exceptions the results from water sample to water sample were fairly consistent. The Mediterranean had a higher saltness than the Atlantic which in turn was more salty than the English Channel. Note the number of salts determined. All of these samples were taken at the surface and away from the land. The analytical procedure was similar to that done before. Qualitative tests were as before and quantitatively the method used was that of evaporation. Like Lichtenberg previously, Bouillon-Lagrange and Vogel measured the carbon dioxide ("gaz carbonique") by heating the water sample and driving the gas into lime water. The residue was filtered and weighed (29, p.502). They arrived at the conclusion that there was no sodium sulfate or calcium chloride in sea water as others had said (29, p.514).

To Bouillon-Lagrange and Vogel the results indicated that the Atlantic and the English Channel contained more carbon dioxide than the Mediterranean. This was attributed to the lower temperature of the Atlantic and the Channel (29, p.514).[118] The deliquescense of the salt mixture obtained from the evaporation of sea water was finally clearly explained here as due to the presence of deliquescent "magnésie muriate" ($MgCl_2$).

> Que le muriate de magnésie est le seul sel deliquescent existant dans l'eau de mer, d'oú provient la propriété qu'a le muriate de soude impur de s'humecter au contact de l'air. (29, p.514)

The knowledge of a number of excellent, very sensitive precipitating agents did exist in the last quarter of the eighteenth century, and Bergman had summarized most of them. Even though these tests were potentially of quantitative use due to their sensitivity there was virtually no attempt to use them as such in mineral water analysis. Mineral water analyses like the sea water analyses given above often did not agree. The analyses of Vogel and Bouillon-Lagrange illustrate the widely varying data for sea water analyses existing at that time.[119]

These analyses of sea water, like those of mineral waters, followed Bergman's recommendations and used evaporation and solvent extraction as the quantitative means. The discrepancies in the results were either due to faults in the analytical procedure or the nature of the sea water itself or both. While most chemists seemed to believe in some consistency of sea water there was no real experimental evidence. It should be remembered that there had not yet been a particularly large number of analyses and little attempt at consistency by analysis of the same waters. The fault lay primarily in the evaporation-solvent-extraction method. Although Boyle had experienced difficulties with the technique as the primary analytical procedure to determine total salt in sea water over 100 years prior, evaporation was still the basic tool in water analysis. The method had been tremendously refined by the use of solvent extraction, but even with this and the solvents available at that time, the method as espoused by Bergman and used in these sea water analyses above was not capable of giving an accurate description of the salt content in sea water. Nevertheless the means for accurate analyses of sea water were known. This was by the use of precipitating agents used hitherto as qualitative tests. Klaproth had visualized their use but he failed to carry this further in his analysis of sea water (see p.55).

MURRAY

Without question the most detailed and definitive work of a chemical nature that had been done on sea water to this date appeared as two lengthy articles in the *Transactions of the Royal Society of Edinburgh* for the year 1818. This work was written by the English chemist John Murray (d. 1820) and had been read to the Society two years earlier. The first article was a detailed analysis of sea water (226). The second was a general formula for the analyses of mineral waters (227). The chemistry and chemical techniques used in both are virtually identical. The general analysis was essentially an application of the methods the author had previously used in the analysis of sea water to the wider field of water analysis in general. Since both papers were read within two months of one another it seems apparent that the formulation of a general analysis method was not an afterthought but rather a primary intent. In at least one place in the paper on sea water Murray made reference to his previously existing general formula (226, p.207), although it had not yet been either published or read.

The sea water analysis seems to have been simply a major application of the method probably prompted because of the discordant results in the sea water analyses in the literature. Murray had noticed that the composition of sea water as reported by various workers differed appreciably in their qualitative and quantitative approach (226, p.205). He carefully examined the previous attempts to analyze sea water and tried to explain all of the possible discrepancies. In fact in many cases he repeated the analysis exactly as it had been done by prior chemists (226, pp.208–216). Upon completion of these analyses it was evident to Murray that previous methods of analyses were of sufficient uncertainty to justify a new method. He showed, for example, in an exhaustive duplication of Lavoisier's analyses of sea water, that the salt

TABLE VII

The results of a sea water analysis by Murray according to the method of Lavoisier

By this analysis, then, the substances obtained from 4 pints of sea water, and their proportions, are as follows:

Muriate of soda		728.5 grains
Muriate of magnesia real	83.5 grains	
	7.9	
	6.2	
	99.4	
Sulphate of magnesia crystallised	6.3	
	8.9	
	33	
	48.2, or real,	23.5
Sulphate of soda crystallised	18	
	44.2	
	6	
	68.2, or real,	30.2
Sulphate of lime, real	22	
	3	
	25	25
Carbonate of lime	1.2	
	1.3	
	2.5	2.5
Carbonate of magnesia	1.7	
	2.8	
	4.5	4.5

(226, pp.214–215)

fractions obtained by evaporation and solvent extraction were by no means pure (226, pp.213–215). In the duplication of Lavoisier's method Murray obtained the results shown in Table VII. He felt that carbonate of lime and carbonate of magnesia were not original components of sea water, but arose from the decomposition of muriate of lime ($CaCl_2$) and muriate of magnesia ($MgCl_2$) and the subsequent absorption of carbonic acid (carbon dioxide) from the air during evaporation (226, p.215). Special tests tried for these two salts identified neither carbonate. Murray then converted the results to those shown in Table VIII.

Next Murray chose to perform another analysis on the same volume (four pints) of sea water from the same locale. This method of analysis was that which was in common usage at the time and was a direct one based more on

TABLE VIII

Conversion of the results by Murray given in Table VII

With these corrections, and reducing the proportions to a pint, the ingredients and their quantities will be as follows:

muriate of soda	182.1 grains
muriate of magnesia	25.9
sulphate of soda	7.5
sulphate of magnesia	5.9
sulphate of lime	7.1
	228.5

(226, p.215)[120]

TABLE IX

Results of a sea water analysis by Murray based on an evaporation and crystallization method

muriate of soda	184 grains
muriate of magnesia	21.5
sulphate of magnesia	12.8
sulphate of soda	2.
sulphate of lime	7.3
	227.6

(226, p.221)

evaporation and crystallization rather than the solvent action of alcohol of Lavoisier's method. The results for one pint are shown in Table IX. While the total salt content obtained did not disagree too badly (228.5 versus 227.6 grains) and though the salts determined were the same, the relative amounts with the exception of muriate of soda and sulphate of lime varied appreciably. Murray felt that the primary source of the difference was the nature of the solvent action of alcohol in reforming some of the sulphate of soda that was decomposed earlier in the salt mixture (226, pp.222–223). Murray determined that if muriate of magnesia and sulphate of soda were dissolved in water and the resulting solution reduced one would find sulphate of magnesia and murate of soda in the resulting solids. If, however, these two original salts were dissolved with the aid of heat in alcohol, they can be recovered as such (226, pp.224–225).

The earlier analyses of mineral waters had also been discordant in their results because of the variety of methods used and the method often involved subjective errors. For this reason he proposed a general formula for the analysis of all waters (227). Two methods for water analyses had been in common use: one by evaporation and the other by precipitation, which Murray termed the direct and indirect methods:

> . . . the *direct method*, in which, by evaporation, aided by the subsequent application of solvents, or sometimes by precipitants, certain compound salts are obtained; and what may be called the *indirect method*, in which, by the use of re-agents, the principles of these salts, that is, the acids and bases of which they are formed are discovered, and their quantities estimated, whence the particular salts, and their proportions, may be inferred. (227, p.260)

Previously the indirect method had been used primarily as a series of qualitative tests prefacing generally the determination of salts weight by the former method.

Murray's general formula was in reality simply an attempt to implement the use of the indirect rather than the direct. His comments supporting such a change were very lucid and clear.

> Chemists have always considered the former of these methods as affording the most certain and essential information: they have not neglected the latter; but they have usually employed it as subordinate to the other. The salts procured by evaporation, have been uniformly considered as the real ingredients, and nothing more was required, therefore, it was imagined for the accuracy of the analysis, than the obtaining them pure, and estimating their quantities with precision. On the contrary, in obtaining the elements merely, no information, it was believed, was gained with regard to the real composition, for it still remained to be determined, in what mode they were combined, and this, it was supposed, could be inferred only from the compounds actually obtained. This method, therefore, when employed with a view to estimate quantities, has been had recourse to only to obviate particular difficulties attending the execution of the other, or to give greater accuracy to the proportions, or, at farthest, when the composition is very simple, consisting chiefly of one genus of salts.
>
> Another circumstance contributed to lead to a preference of the direct mode of analysis; the uncertainty attending the determination of the proportions of the elements of compound salts. This uncertainty was such, that even from the most exact determination of the absolute quantities of the acids and bases existing in a mineral water, it would have been difficult, or nearly impracticable, to assign the precise composition, and the real proportions of the compound salts; and hence the necessity of employing the direct method of obtaining them.
>
> (227, p.260)

Murray felt that the level of chemistry at that time was such that the composition of the components could be determined (227, p.260). More important the composition of a water, he felt, could be determined with greater accuracy by the indirect method (227, pp.261–262), and the indirect, or precipitation method was, to Murray, easier to perform (227, p.262).

The results obtained in a direct analysis, Murray pointed out, did not necessarily represent the real or actual salts in the water.

> Another advantage is derived from these views, if they are just, that of precluding the discussion of questions which otherwise fail to be considered, and which must often be of difficult determination, if they are

> even capable of being determined. From the state of combination being liable to be influenced by evaporation, or any other analytic operation by which the salts existing in a mineral water are attempted to be procured, discordant results will often be obtained, according to the methods employed; the proportions at least will be different, and sometimes even products will be found by one method which are not by another. (227, p.263)

This idea was not original to Murray but what was novel was that he did give experimental proof, conclusive proof, as to the inadequacy of the direct method.

The indirect method as proposed by Murray was simply the determination by precipitation of the specific "acids and bases" in the water sample[121] rather than the removal of the salts per se.

> In a water which is of complicated composition, this will more peculiarly be the case. The Cheltenham waters, for example, have, in different analyses, afforded results considerably different; and, on the supposition of the salts procured being the real ingredients, this diversity must be ascribed to inaccuracy, and ample room for discussion with regard to this is introduced. In like manner, it has often been a subject of controversy, whether sea water contains sulphate of soda with sulphate of magnesia. All such discussions, however, are superfluous. The salts procured are not necessarily the real ingredients, but in part, at least, are products of the operation, liable, therefore, to be obtained or not, or to be obtained in different proportions, according to the method employed. And all that can be done with precision, is to estimate the elements, and then to exhibit their binary combinations according to whatever may be the most probable view of the real composition. (227, p.263)

The scheme, with the subsequent necessary washings, dryings and weighings, was essentially this. After preliminary tests as to probable composition,

> the presence of sulphuric and carbonic acids being detected by nitrate of barytes, of muriatic acid by nitrate of silver, of lime by oxalic acid, of magnesia by lime-water or ammonia, and of any alkaline neutral salt by evaporation. (227, p.264)

The water was evaporated slowly as far as possible without any noticeable precipitation or crystallization of muriate of barytes (barium chloride) was added to precipitate sulfuric acid (sulfate ion).[122] Lime (calcium oxide) was precipitated as its respective oxalate by the addition of oxalate of ammonia and then converted to the sulfate to be weighed. The method for the precipitation of magnesia[123] was accomplished by the precipitation from solution as the phosphate of ammonia, and magnesia by the addition of carbonate of ammonia and phosphate of soda (or phosphoric acid as the case may be) and weighed as the phosphate (actually the pyrophosphate) of magnesia. The muriate of soda (sodium chloride) was determined from the residual liquors, the muriatic acid being inferred from this amount (227, p.273), 100 grains of muriatic acid being equivalent to 53.3 of soda and 46.7 of the acid. If the muriatic acid present was greater or less than that in the muriate of soda then the muriatic acid would be determined by precipitation with nitrate of silver or nitrate of lead (227, pp.267–274):

> The real quantity will thus be determined with perfect precision, and the result will form a check on the other steps of the analysis, as it will lead to the detection of any error in the estimate of the other ingredients. (227, p.274)

And:

> Thus, by these methods, the different acids, and the different bases are discovered, and their quantities determined. To complete the analysis, it remains to infer the state of combination in which they exist. It will probably be admitted, that this must be done on a different principle from that on which the composition of mineral waters has hitherto been inferred. The compounds which may be obtained by direct analysis, cannot be considered as being necessarily the real ingredients, and to state them as such would often convey a wrong idea of the real composition. (227, pp.274–275)

In conclusion Murray stated:

> The results of the analysis of a mineral water may always be stated, then, in these three modes: *1st*, The quantities of the acids and bases: *2dly*, The quantities of the binary compounds, as inferred from the principle, that the most soluble compounds are the ingredients; which will have at the same time the advantage of exhibiting the most active composition which can be assigned, and hence of best accounting for any medicinal powers the water may possess: And, *3dly*, The quantities of the binary compounds, such as they are obtained by evaporation, or any other direct analytic operation. The results will thus be presented under every point of view. (227, p.275)

To Murray there was a real question as to whether binary compounds existed at all in a state of solution (226, pp.227–228). If this were so it was evident that muriate of soda must be the chief component in sea water, although its quantity might vary with interactions with other salts (226, p.228). Similarly muriate of magnesia should exist in large amounts, its actual proportions also subject to change by interaction.

The principal difficulty is with regard to the sulphate of magnesia, and the sulphate of soda. It has always been supposed, that sulphate of magnesia is an ingredient in sea water, from its being procured by evaporation; and it is possible that it may be so. But it is just as possible, a priori, that sulphate of soda may be the original ingredient, and that, during the evaporation, the mutual action between it and muriate of magnesia, is favoured by the concentration, whence portions of both are decomposed, and corresponding quantities of sulphate of magnesia and muriate of soda are formed. Nor is there any thing connected with the mere results themselves, which proves which of these views is just.

If the appeal be made to experiment, it is sufficiently established, that sulphate of magnesia may be formed by the action of sulphate of soda on muriate of magnesia. When these two salts are boiled together in solution, a double decomposition takes place at least partially, and portions of sulphate of magnesia and muriate of soda are formed.

(226, p.228)

If, then, one accepted the view that the sulphate existing in sea water was sulphate of magnesia rather than sulphate of soda then the first and second analyses would have the values shown in Table X. The second one is confusing and was even more difficult to support. In other words, Murray felt that any separation of component by crystallization, no matter what the method, was inaccurate. As a consequence, Murray employed the indirect method outlined above, especially since it made little difference in this method whether the constituents did or did not exist as binary compounds. The results are shown in Table XI. Murray then presented the possible ways

TABLE X

Results of the analysis by Murray of sea water with the assumption that the sulphate exists as sulphate of magnesia

For the first:	
muriate of soda	188.3 grains
muriate of magnesia	16
muriate of lime	5.8
sulphate of magnesia	18.4
	228.5
and the second:	
muriate of soda	185.6
muriate of magnesia	15.2
muriate of lime	5.9
sulphate of magnesia	20.9
	227.6

(226, p.233)

TABLE XI

The results of sea water analysis by Murray using the indirect method

The elements, then, of the salts, in a pint of sea water, are, by this analysis:

lime	2.9 grains
magnesia	14.8
soda	96.3
sulphuric acid	14.4
muriatic acid	97.7
	226.1

(226, p.237)[124]

these "acids and bases" could exist in sea water:

> The proportions of the compound salts may be assigned from these, according to whatever view may appear most probable, of the state of combination in which they exist in sea water, and thus the results may be compared with those of the former analyses.
> Thus, supposing the elements to be combined in the modes in which they are obtained by evaporation, that is, as muriate of soda, muriate of magnesia, sulphate of magnesia, and sulphate of lime; the proportions of these salts in a pint, will be: (226, p.238)

muriate of soda	180.5 grains
muriate of magnesia	28
sulphate of magnesia	15.5
sulphate of lime	7.1
	226.1

If he assumed that the lime existed as the muriate, which he felt was most probable, and that the sulphuric acid existed as sulphate of magnesia, then the proportions would be as shown in Table XII, or if the sulphuric acid existed as the sulphate of soda as shown in Table XIII.

TABLE XII

The results of Murray's sea water analysis with the assumption the lime existed as muriate

muriate of soda	180.5 grains
muriate of magnesia	18.3
muriate of lime	5.7
sulphate of magnesia	21.6
	226.1

(226, p.238)

TABLE XIII

The results of Murray's sea water analysis with the assumption that the sulphuric acid existed as a sulphate of soda

muriate of soda	159.3 grains
muriate of magnesia	35.5
muriate of lime	5.7
sulphate of soda	25.6
	226.1

(226, p.238)

> These proportions differ somewhat, though not very materially, from those found by the other modes of analysis. The principal differences consist in the quantity of magnesia, and of sulphuric acid being rather larger. This is evidently to be ascribed to the modes of detecting sulphuric acid by barytes, and magnesia by phosphoric acid and ammonia, being so perfect, that the entire quantities of them are found; while, in the other modes, from the difficulty of effecting the entire separation of salts from each other, a small portion of sulphate of magnesia, or of muriate of magnesia and sulphate of soda, had remained with the muriate of soda, and though subcarbonate of soda was employed to decompose them, this decomposition is not altogether perfect. (226, pp.238–239)

Murray tended to believe that the sulphuric acid existed as sulphate of soda. For that reason he regarded the last of these three inferences to be the best approximation of the real composition of sea water, both in ingredients and proportions thereof.

Although Murray did believe in the superiority of the analysis on the basis of acids and bases obtained, he was aware that the inference as to the binary compounds present was somewhat subjective. For this reason he stated:

> Of the different views which may be taken of the state of combination of the elements, I have already inferred, that the one which supposes the sulphuric acid to exist in the state of sulphate of soda, is the most probable; and as the mode of analysis by re-agents is the most accurate, the last table may be considered as that which exhibits the highest approximation to the real composition of sea water, both with regard to its ingredients, and their proportions. (226, p.240)

As Murray pointed out, prior mineral water analyses by other chemists had yielded erratic results (227, p.259). Sea water analyses had also left much to be desired in terms of consistency. Sea water had always been considered and treated simply as a mineral water, although perhaps just a more concentrated one. There is no indication that Murray felt differently. In fact Murray's interest in the sea seems to have been prompted purely for academic reasons in establishing a general procedure for water analysis rather than by any special interest in the sea. He appears to have chosen sea water as the means to drive home the value of the indirect method since no mineral water was as concentrated or presented as many problems as sea water. In this regard *this is the first time sea water analyses have been of specific value to water analysis*. Previously the reverse had been the case. Sea water analysis now became important to mineral water analysis because due to its concentration the inadequacies of the methods of water analysis readily became more apparent.

Murray was not particularly interested in the sea and its study. Rather he was interested only in the analysis of sea water as a difficult, representative, water sample. Implicit in his work was the belief that sea water was constant for he apparently assumed that samples from a variety of locales could be directly compared. He was careful to choose water samples far away, he thought, from rivers, so as to be representative of the sea.[125] He ran numerous analyses by different methods on exactly the same water sample. With Murray's work there existed reproducibility and consistency in the results of sea water analyses.

Murray's work was entirely gravimetric. His technique even from a modern standpoint was excellent, so much so that the four quantitative determinations he recommended have existed until the present time with only minor modifications.[126]

The concept of determining the "acids and bases" present in a sample was to become extremely important in analytical chemistry. The work of John Murray was largely responsible for this. It gave chemists a general method which when used with care was capable of giving reproducible results in mineral water as well as sea water analyses.

The foregoing discussion on the work of Murray may have left the impression that he was solely responsible for the chemistry in the general method for water analysis that he proposed. Such is not the case. With the exception of the magnesia determination by Murray (see note 123), all of the chemistry existed at that time (227, p.278). Murray did, of course, blend it into a precise, reproducible scheme with the emphasis being on the fineness of the analysis being expressed in "acid and base" content. This is given here not to lessen Murray's contribution, but rather to bring things into proper perspective.

MARCET

Another person who in this time period contributed significantly to the study of the sea was the English chemist Alexander Marcet (1770–1822), a Fellow of the Royal Society. As a medical doctor, much of Marcet's work was oriented toward the medical (240, vol.3, p.707). Nevertheless, he was a chemist, and between the years 1807 and 1822 wrote several papers on the sea and its water (199,200,201,202,203).[127] The first of these was not specifically on ocean water but rather "An Analysis of the Waters of the Dead Sea and the River Jordan" (199).[128] According to Marcet, he undertook the analysis of these waters for this reason:

> The names of Lavoisier, and of his two distinguished associates, might appear to render any further investigation of the nature of this water superfluous; but whoever has perused the paper in question, must be convinced, that these gentlemen, however correct in their general statements, neither attained that degree of accuracy of which modern analysis is susceptible, nor did they bestow on the subject that share of attention which is indispensable in minute analytical experiments.
>
> (199, p.297)[129]

The most interesting thing about his paper on the Dead Sea waters is that entirely independent of Murray (at that time), Marcet devised a scheme of analysis that was essentially the same as Murray's.[130] By the time Marcet had written his major paper on sea water (210), he was familiar with what Murray had said and done in this area.

It is satisfactory to observe that Dr. Murray adopted, several years afterwards, from considerations of the same kind, a mode of proceeding precisely similar, and indeed that he proposed in a subsequent paper, a general formula for the analysis of mineral waters, in which this method is pointed out as likely to lead to the most accurate results. And this coincidence is the more remarkable, as it would appear, from Dr. Murray not mentioning my labours, that they had not at that time come to his knowledge. (201, p.194)

The primary exceptions in method were determinations of magnesia and chlorine (muriatic acid). Marcet relied on the direct action of muriatic acid (hydrochloric acid) by silver nitrate precipitating *luna cornea* (AgCl) which was dried, weighed, heated to decomposition, and the silver collected, weighed and the acid calculated (199, pp.301–302). Marcet reported the analysis results as the acid and the salts (see Table XIV). However, he preferred to look at it in terms of the salts he thought there (Table XV).

TABLE XIV

The results of Marcet's analysis of Dead Sea water

250 grains of the Dead Sea water appear to contain:	salts	acid
muriate of lime	9.480 grains	4.66 grains
muriate of magnesia	25.25	14.15
muriate of soda	26.695	12.28
sulphate of lime	0.136	–
	61.561	31.09

(199, p.309)

TABLE XV

Marcet's interpretation of the salts contained in Dead Sea water

And therefore 100 grains of the same water would contain:	
muriate of lime	3.792 grains
muriate of magnesia	10.100
muriate of soda	10.676
sulphate of lime	0.054
	24.622

(199, p.309)[131]

Twelve years later in 1819, Marcet wrote another paper on sea water: "On the specific gravity, and temperature of sea waters, in different parts of the Ocean, and in particular seas; with some account of their saline contents" (201). As his reason for this second article, he explained that "while analyzing the waters of the Dead Sea and the River Jordan, about twelve years ago, it occurred to us that a chemical examination of different seas, in a variety of latitudes and at different depths, might be interesting" (201, p.161). The fairly long time span between these two papers was due to a number of things. Since he did not go to sea, Marcet had to rely on the provision of water samples by friends (201, p.161). By the time he had a large variety of such samples several years had passed. Then an accident (240, vol.3, p.703) caused the untimely death of Marcet's friend Tennant in 1815. As Marcet explained:

> Procrastination and delay were the natural consequence of this misfortune; and I should probably have entirely lost sight of the subject, had not my intention been again directed to it by the late expeditions to the Arctic regions, and the great zeal and kindness of some of the officers engaged in them, in procuring for me specimens of sea water, collected in different latitudes, and under peculiar circumstances, so as to add greatly to the value of those which I previously possessed.
>
> (201, p.162)

This lengthy paper of Marcet's virtually abounds in information. Furthermore, it is without question the most definitive work thus far (by 1819) on density, temperature, and saline content. Marcet determined the density of 68 water samples from a number of seas and included the first detailed discussion of water sampling apparatus (see Appendix II).[132]

Marcet's paper was divided into two parts. The first dealt primarily with the specific gravities of sea water from voyaging seas (201, pp.161–190). The second dealt with the saline contents of these waters (pp.191–201). In the second part of his paper Marcet chose to analyze a "selected" number of his water samples as being representative of a number of different seas and bearings.

> An accurate analysis of all the specimens which I have noticed in this paper would have been a most laborious, and indeed almost interminable undertaking, which would not have afforded any adequate object of curiosity or interest. All that I aimed at, therefore, was to operate upon a few of the specimens, so selected as to afford a general comparison between the waters of the ocean in distant latitudes and in both hemispheres, and to enable me also to ascertain whether particular seas differed materially in the composition of their waters.[133]
>
> (201, p.191)

As Marcet summarized his method of analysis, the procedure was:

> 1st. To ascertain the quantity of saline matter contained in a known weight of the water under examination, desiccated in a uniform and well defined mode; and to compare it with the specific gravity of the water.
>
> 2ndly. To precipitate the muriatic acid from a known weight of the water, by nitrate of silver.
>
> 3dly. To precipitate the sulphuric acid by nitrate of barytes, from another similar portion of water.
>
> 4thly. To precipitate the lime from another portion of water, by oxalate of ammonia.
>
> 5thly. To precipitate the magnesia from the clear liquor remaining after

the separation of the lime, which is best effected by phosphate of ammonia, or of soda, with the addition of carbonate of ammonia.

The soda, by this method, is the only ingredient which is not precipitated, and which, therefore, can only be inferred by calculation. (201, pp.193–194)

The method used in analysis was essentially the same Murray had used. There was some question on Marcet's part as to the manner in which the sulfuric acid (sulfate ion) was combined. Marcet also preferred to precipitate the muriatic acid and calculate the soda content (NaO). His results for 500 grains of sea water, sample No. 27 from the middle of the North Atlantic (201, p.196), expressed as the acids and bases were as shown in Table XVI. In the supposed state of combination the values were (Table XVII):

TABLE XVI

The analysis results by Marcet of sea water from the North Atlantic given as acids and bases

Muriatic acid	8 grains
Sulphuric acid	1.27
Lime	0.314
Magnesia	1.08
Soda	8.11
	18.774

(201, p.198)

TABLE XVII

Marcet's inferred state of combination of the results of Table XVI

Muriate of soda	13.3 grains
Sulphate of soda	2.33
Muriate of lime	0.616
Muriate of magnesia	2.577
	18.823

(201, p.198)

By drying a sample of muriate of lime and one of muriate of magnesia at 212°F, Marcet found that they lost water. He then corrected the values above. These then became (Table XVIII):

TABLE XVIII

The corrected values for Table XVII after drying

Muriate of soda	13.3 grains
Sulphate of soda	2.33
Muriate of lime	0.975
Muriate of magnesia	4.955
	21.460

(201, p.199)[134]

Of the 15 samples plus that of the Dead Sea that Marcet analyzed, only No.27 was expressed in the detail above. This was probably because he felt that this water was representative of the oceans. For the remaining samples the results (see Appendix III) were given as the dried weight of precipitated "acid or base."[135]

Toward the end of the paper Marcet threw out the conjecture that sea water might contain a large number of substances:

> I, in my turn, put the question to Dr. Wollaston whether it was not probable that minute quantities of all soluble substances in nature might be detected in sea water? (201, p.200)

This conjecture was in partial answer to a question of Dr. William Hyde Wollaston (1766–1828) directed to his close friend Marcet, as to whether or not sea water contained any potash. At Marcet's insistence, and having supplied him with samples, Wollaston soon communicated his discovery of potash in sea water in a letter to Marcet:

> The expectation which I expressed to you, that potash would be found in sea water, as an ingredient brought down by rivers from the decay of land plants, is now fully confirmed by experiments on waters obtained from situations so remote from each other, as to establish its universality.
>
> There is no difficulty in proving the presence of this ingredient by muriate of platina. For though the triple muriate of platina and potash is so soluble that this reagent causes no precipitate from sea water in its ordinary state, yet when the water has been reduced by evaporation to about 1/8th part, so that the common salt is beginning to separate by crystallization, the muriate of platina then causes a copious precipitate. (319, p.325)

The amount of potash Wollaston had stated present was: "since the pint of water weighed about 7520 grains, $\frac{6.4}{7520}$ gives the proportion of potash, about $\frac{1}{1200}$; but the quantity of mere potash is less than $\frac{1}{2000}$th part of sea water, as its average density" (319, p.326). As Marcet had pointed out: "Dr. Wollaston thinks it probable that the potash exists in sea water in the state of sulphate" (201, p.201).

This conjecture of Marcet's that the sea might contain all soluble substances in nature is not unreasonable in that the substances in the sea were by then generally considered to have come from the land. Yet no one prior to Marcet appears to have explicitly stated it.

> For the ocean having communication with every part of the earth through the rivers, all of which ultimately pour their waters into it; and soluble substances, even such as are theoretically incompatible with each other, being almost in every instance capable of co-existing in solution, provided quantities be very minute, I could see no reason why the ocean should not be a general receptacle of all bodies which can be held in solution. (203, p.448)

This question in general and an earlier article by the French chemist G.F. Rouelle (1703–1770) that had just come to Marcet's attention describing the appearance of mercury in sea salt prompted Marcet to other research which culminated in the writing of another paper on sea water (202) published in 1820. In spite of very detailed and precise work, Marcet did not discover mercury in sea water.[136] He rather charitably stated that he was "justified in concluding that the mercury, which other chemists have detected in sea-salt or its products, must have been introduced there from some local or accidental circumstances" (202, p.451).

Marcet had also tried to find some nitric salt (a nitrate) in sea water as well as alumnia, silica, iron, copper, and sal ammoniac (NH_4Cl). Of these he was only able to find the sal ammoniac.[137] He identified carbonate of lime in solution[138] as well as a triple sulphate of magnesia and potash (202, p.456). Marcet said that sea water contained no muriate of lime ($CaCl_2$). He added:

> Some of these circumstances [identification of new substances] will, of course, require that former analyses of sea water, and my own in particular, should be corrected and revised; but this I shall not attempt to do, until I have obtained farther, and still more precise information on the subject. (202, p.456)

Marcet never resolved a number of these questions, especially that of the presence of muriate of lime. He died within the year.

Apparently he never attempted to offer any explanation for the absence of mercury and other substances, although he originally thought they might exist in sea water. He knew that sea water contained lime (CaO, or Ca^{2+}) as well as muriatic acid (HCl, or Cl^-), yet in these last analyses he did not find muriate of lime ($CaCl_2$). This last short paper did not tell much about the methods used as a basis for these conclusions – nothing, for example, about the muriate in lime trials. It is difficult to say why he did not find this salt. He should have.

Toward the end of his life, Marcet seemed to revert more in the direction of expressing the content of these waters in terms of the salts he thought present rather than the components (acids and bases).[139]

The analysis by Marcet on the 14 sea water samples from the Arctic, North and South Atlantic, White Sea, Baltic Sea, Sea of Marmora, Yellow Sea and Mediterranean represented the first attempt to compare waters from a large number of ocean bodies.

The works of Murray and Marcet were well read and thus it might seem that the emphasis on the use of precipitation methods to determine the composition of salt in sea water would result in the falling of evaporation procedures from popular usage. And such was the case (16, 114, 115, 173, 244).

TABLE XIX

The results of sea water analysis by Friedman Göbel

Nach	1. Schwefelsäure	1.3478 Gramm
„	3. Chlor	15.5421
„	4. Kali	0.1390
„	4. Natron	6.7900
„	4. Talkerde	4.8300
„	4. Brom	0.0061

(115, p.14)

In the time period between 1820 and 1865 there were numerous analyses of sea water most of which followed Murray's method. For example, Dr. Friedman Göbel (114, 115) determined the results given in Table XIX. The method of determination was essentially that of Murray. As was usually the case, he inferred the composition of these "acids and bases" in the sea water as shown in Table XX. He added, "Kalk, Lithion, Jod, Kohlensäure und phosporsäure Salze, so wie erzmetallische Bestandtheile, sind im Eltonwasser nicht vorhanden" (115, p.15). Göbel analyzed other seas, specifically the Caspian (115, p.107). He evidently used only surface samples.

The German chemist Ernst von Bibra published in 1851 an analysis of waters from the Pacific and Atlantic (16). He analyzed three samples from the Pacific, five from the Atlantic and one from the North Sea (see Table XXI). He used Murray's recommended method.

The noted French chemist Louis Nicolas Vauquelin (1763–1829) in 1825 made a not overly successful attempt at the analysis of some of the salts in sea water by their precipitation with soap (296). E.J.B. Boussingault performed a number of salt and fresh water analyses (32, 33, 34, 35). The method he strongly recommended and used was that of Murray.

The German M.D. and chemist Christian Heinrich Pfaff (1773–1852),

TABLE XX

The inferred state of composition of acids and bases of Table XIX

Hundert Theile des Wassers vom Elton-See enthalten hiernach:	
1.665	Schwefelsäure Talkerde
0.222	Chlorkalium
13.124	Chlorsodium
10.542	Chlortalcium
0.007	Bromtalcium
74.440	Wasser nebst unwegbaren Quantitäten organischer Substanzen
100.000	Gramm

(115, p.15)

TABLE XXI

The analysis of sea water by Ernst von Bibra

	I	*II*	*III*	*IV*	*V*
Chlornatrium	2.4825	2.8391	2.5885	2.5877	2.6333
Bromnatrium	0.0402	0.0441	0.0208	0.0401	0.0420
Schwefels. Kali	0.1409	0.1599	0.1418	0.1359	0.1327
Schwefels. Kalk	0.1488	0.1449	0.1622	0.1622	0.1802
Schwefels. Talkerde	0.0947	0.1041	0.1117	0.1104	0.1079
Chlormagnesium	0.3681	0.3852	0.4884	0.4345	0.3802
	3.2752	3.6773	3.5233	3.4708	3.4763
Wasser	96.7248	96.3227	96.4767	96.5292	96.5237
Gramm	100.0000	100.0000	100.0000	100.0000	100.0000

Chlor wurde zu wenig gefunden in I: 0.0096; in III: 0.0102; in IV: 0.0098

	VI	*VII*	*VIII*	*IX*	*X*
Chlornatrium	2.7558	2.7892	2.6424	2.9544	2.5513
Bromnatrium	0.0326	0.0520	0.0400	0.0500	0.0373
Schwefels. Kali	0.1715	0.1810	0.1625	0.1499	0.1529
Schwefels. Kalk	0.2046	0.1557	0.1597	0.1897	0.1622
Schwefels. Talkerde	0.0614	0.0584	0.0778	0.1066	0.0706
Chlormagnesium	0.0326	0.3332	0.4022	0.3916	0.4641
	3.2585	3.5695	3.4746	3.8422	3.4383
Wasser	96.7415	96.4305	96.5254	96.1578	96.5017
Gramm	100.0000	100.0000	100.0000	100.0000	100.0000

Chlor wurde zu wenig gefunden in VIII: 0.0110, und in X: 0.0087

(16, p.98)

who wrote in 1821 an extensive (general) and comprehensive chemical textbook (283, p.154), *Handbuch der analytischen Chemie,* performed a large number of analyses on sea waters (244, 245, 247, 248, 249) from 1814 to 1825. Pfaff, working at the University of Kiel, part of which later became that great oceanographic center, was an extremely accurate and conscientious worker. Upon becoming aware of Murray's and Marcet's work in this field, he immediately began new analyses, most of which dealt with Baltic Sea Waters. Pfaff was convinced of the preferability of the reporting of the acids and bases present. He clearly pointed out that the compounds (salts) were formed during the evaporation and residue formation.

> Es ergiebt sich aus dieser neuen Analyse des Osteewassers, dass die Bestandtheile desselben, wenn man von ihrer Verbindung zu Salzen hierbei abstrahiert, folgende sind; von Säuren: Kohlen-säure, Schwefelsäure, Salzsäure, Hydroiodsäure; von Basen: Eisenoxyd, Talkerde, Kalk,

> Natron, Kali, Ammoniak, und dass die Verbindungen, die sich beim allmäligen Abrauchen bilden, und die Ausscheidungen hierbei, in folgender Ordnung nach einander sich zeigen: Eisenoxyd, Kohlensäurer Kalk, Schwefelsäurer Kalk, Kochsalz, Bittersalz, ein Doppelsalz aus Schwefelsäure, Kali und Talkerde; und dass zuletzt eine unkrystallisierbare Mutterlauge zurückbleibt, welche zweierlei Arten von Extractivstoff, ein hydroiodsäures Salz (hydroiodsäure Talkerde?) und salzsäure Talkerde enthalt. (249, pp.381–382)

He reported the acids and bases as well as probable compounds formed. Like Marcet, he was unable to determine any mercury in sea water.

G. Laurens, who detected iodine in sea water, analyzed Mediterranean waters (173). The results are shown in Table XXII. Laurens was familiar with

TABLE XXII

The analysis of Mediterranean Sea water by G. Laurens

En résumé, 100 parties d'eau de la mer Méditerranée, prise à deux lieues environ de distance du port de Marseille, et à la surface, m'ont donné:

chlorure de sodium	2,722
chlorure de magnésium	0,614
sulfate de magnésie	0,702
sulfate de chaux	0,015
carbonate de chaux	0,009
carbonate de magnésie	0,011
acide carbonique	0,020
potasse	0,001
matière extractive	les traces
iode	quant. indét.

(173, p.92)

Marcet's and Wollaston's work and determined potassium using Wollaston's (319) method (173, p.91). In determining the total solids, he initially evaporated the sample to dryness and heated further to a dark red.

Yet Murray's method had not swept the field; it was still common for sea water analyses to be done using the evaporation techniques.

Dr. Andrew Fyfe examined in 1819 a variety of waters from the North Polar Sea in an attempt to determine the total gravity of saline matter as well as to compare with other waters analyzed previously by other chemists (101). He used specific gravity (weighed) and evaporation techniques. In 1838, a French chemist, Darondeau, examined a total of ten samples, each taken from five locales at surface and at depth, from the world's major oceans (66). He was aided in his analyses by the famous chemist Theophile

TABLE XXIII

The analysis of sea water by Darondeau

Epoques auxquelles l'eau a été prise, et lieux d'où elle provient	*Latitude*	*Longitude*	*Profondeurs auxquelles l'eau a été prise*	*Densité à 80 et 10° cent.*	*Résidus salins pour 100 part. d'eau*	*Quantités de gaz pour 100 part. d'eau à 0° de température. et 700 mm de pression*	*Composition de 100 parties du gaz/gram*		
							oxigène	*azote*	*acide carbon.*
50 août 1836	11°8'N	108°50'O	surface	1,02594	3,429	2,09	6,16	83,33	10,51*
Océan pacifique			70 brasses	1,02702	3,528	2,23	10,09	71,05	18,06
19 mars 1837	11 43 N	87 18 E	surface	1,02545	3,218	1,98	5,53	80,50	13,97
Golfe du Bengale			200 brasses	1,02663	3,491	3,04	3,29	38,56	58,15
10 mai 1837	18 0 N	85 32 E	surface	1,02611	3,378	1,91	6,34	80,34	13,32
Golfe du Bengale			300 brasses	1,02586	3,484	2,43	5,72	64,15	30,13
31 juill. 1837	24 5 S	52 0 E	surface	1,02577	3,669	1,85	9,84	77,70	12,46
Océan indien			450 brasses	1,02739	3,518	2,75	9,85	55,23	34,92
24 août 1837 Océan atlantique méridional	30 40 S	11 47 E	400 brasses	1,02708	3,575	2,04	4,17	67,01	28,82

* Dans cette observation, il doit y avoir de l'incertitude sur la quantité d'acide carbonique, parce qu'on ne l'a pas dosée immédiatement.

(66, p.103)

TABLE XXIV

The analysis of North Sea water by G. Clemm

1000 Theile Wasser enthielten:

Chlornatrium	24,84 Gramm, wasserfrei berechnet
Chlormagnesium	2,42
Schwefelsäure Talkerde	2,06
Chlorkalium	1,35
Schwefelsäure Kalkerde	1,20
Kohlensäure Kalkerde Kohlensäure Talkerde Kohlensäures Eisenoxydal Kohlensäures Manganoxydal Phosphorsäure Kalkerde Bromverbindung Jodverbindung Kieselerde Organische Materie Kohlensäure Ammoniak (?)	in sehr geringer, nicht bestimmbarer Menge

55, pp.111–112)

Jules Fremy (1807–1867). The results (see Table XXIII) were determined essentially by an evaporation procedure.

The German chemist G. Clemm in 1841 published an analysis of the waters of the North Sea (55). The results he obtained were as shown in Table XXIV. Clemm, working under the guidance of the great German chemist Wöhler, used an evaporation method. He did not, however, indicate the manner or specific conditions by which he determined the individual components. Presumably it would have to have been similar to Murray's, especially since this chemistry was well known in Germany at this time. In fact, the analytical chemistry that probably began as a Swedish science with Scheele and Berzelius was essentially a German science for much of the nineteenth century (283, p.147). This is especially true for the middle 50 years.

By 1855 the elements that comprise the bulk of the salt matter had been discovered and subsequently detected in sea water. The exception was fluorine presence in sea water, which, although finally detected in a compound in 1865 (100), had been inferred as early as 1850 (313), and was not finally isolated as the element until 1886 (307, p.769). The famous English chemist Sir Humphrey Davy, in 1807–1808 isolated the elements sodium and potassium, as well as barium, strontium, calcium, and magnesium by the electrolysis of the fused bases. The element iodine was discovered in 1811 (57) in the ashes of marine algae and Pfaff, in 1825 detected it in Baltic sea water (249); and it was found in the waters off France ten years later (173).

The first quantitative estimate of the concentration of this element in sea water was not published until 1825 (188). Bromine was discovered in 1824 by the French chemist Antoine-Jerome Balard (1802–1876) by the treatment of waste mother liquor from salt solutions by chlorine (11). Shortly thereafter he identified it in sea water (11). A quantitative estimate in sea water was given in 1838 (115) but it was not until 1871 (287) that modern values were obtained. The element boron was detected by Gay-Lussac and Louis-Jacques Thenard (1777–1857) in 1808 by the treatment of boric acid with potassium (104). The presence of boron in sea water seems first to have been noted in 1853 (258, p.13), but definitely reported in 1865 (100) by Georg Forchhammer, the Danish chemist and geologist, when he reported strontium in sea water.

USIGLIO

In 1849 the Italian chemist J. Usiglio published two articles on the analysis of sea water. In the first of these (249) Usiglio presented the excellent, consistent results of his analyses of the Mediterranean (see Table XXV). In the second paper (295), however, he chose to take issue with evaporational techniques as applied to sea water analysis.

Usiglio undertook all of his analyses in order to try to improve the precision of the previous sea water analyses which he felt left much to be desired (295, p.172). He felt that a number of points in the analysis of sea water could be improved upon. The second of the two series of analyses (295) he undertook because he felt that a more thorough understanding of the contents of the sea and its subsequent evaporation might be of immense practical value to those people who produced salts on an industrial scale by evaporation of Mediterranean waters (295, p.172). Primarily, however, he wanted to show the undesirability of evaporation in sea water analyses (295, p.172). This second memoir dealt almost entirely with analysis by evaporation and subsequent testing.[140]

The chemical methods Usiglio used to determine the constituents in his analyses were those of Murray and Marcet. Usiglio did improve on some of these; however, he used, for example, Wollaston's procedure for the determination of potash (potassium) (319).

Usiglio gave the order of separation of salts from sea water as it was concentrated by evaporation. His results were:

> Cinq litres d'eau de mer pesant 5^{kn}, 129 laissent déposer à l'évaporation les sels suivants:
>
> A 3°,5 de l'aréomètre. Eau de mer ordinaire. Dépôt nul.

TABLE XXV

J. Usiglio's analysis of Mediterranean Sea water

Indication des sels	*Eléments*	*Poids obtenus pour 100 gr. d'eau de mer*		*Poids pour 1 litre d'eau (gr)*	*Observations*	
Oxyde ferrique			0,0003	0,003	sulfate de chaux:	
					sulfate hydraté à 2 équivalents d'eau	9,1716
					et par litre	1,76
Carbonate calcique	acide carbonique	0,0050	0,0114	0,0118		
	chaux	0,0064				
					sulfate de magnésie:	
Sulfate calcique	acide sulfurique	0,0798	0,1357	1,392	sulfate hydraté à 7 équivalents d'eau	0,5051
	chaux	0,0559			et par litre	5,181
Sulfate magnésique	acide sulfurique	0,1635	0,2477	2,541	chlorure magnésique:	
	magnésie				acide chlorhydrique correspondant	0,2441
					magnésie	0,1381
					et par litre: acide chlorhydrique	2,504
Chlorure magnésique	chlore	0,2374	0,3219	3,302	magnésie	1,406
	magnésium	0,0845				
Chlorure potassique	chlore	0,0240			chlorure potassique:	
	potassium	0,0265	0,0505	0,518	potasse correspondante	0,032
					et par litre	0,328
Bromure sodique	brome	0,0432			brome et chlorure sodique:	
	sodium	0,0124	0,0556	0,570	ensemble soude correspondante	1,577
					et par litre	16,177
Chlorure sodique	chlore	1,7854				
	sodium	1,1570	2,9424	30,182		
			3,7655	38,625		
Eau			96,2345	987,175		
Poids total			100,0000	1025,800		

(294, p.104)

A 7° 1,0^{gr}, 336 de carbonate de chaux contenant 0,016 d'oxyde de fer.

De 7° 1 à 14 degrés, des traces d'un sel analogue. Dépôt sensiblement nul.

A 16° 75, un dépôt de 3^{gr}, 065 composé de 0,265 de carbonate de chaux, 2,80 de traces de carbonate de magnésie et de sulfate de chaux hydraté.

A 20° 5, dépôt de 2^{gr}, 81, composé de sulfate de chaux pur, bien défini et cristallisé.

A 22 degrés, dépôt de 0^{gr}, 92 du même sel.

A 25 degrés, dépôt de 0^{gr}, 80 du même sel et cristallisation confuse.

(295, p.175)

Usiglio performed these evaporations himself taking considerable care. These values obtained from evaporation and those obtained in his earlier analyses do not agree.

Usiglio then carried out the partial evaporation of three identical sea water samples to varying specific gravities: 25, 30 and 35 degrees Baumé.[141] The values for density were:

25 degrees density 1.21
30 degrees density 1.264
35 degrees density 1.32

The results shown in Table XXVI were obtained upon the analysis of these three waters.

The results of these three concentrated sea water samples do not even agree in terms of total salt content much less individual components. Even with the extreme care Usiglio used there are differences in the results obtained by the two methods. At increasing densities he showed that the results (Table XXVII) became even more varied (295, pp.186–187).

On the basis of his experiments Usiglio decided that it was much more valid to analyze sea water with no drying or concentration in any way:

> Il a paru convenable de déterminer directement, excepté pour l'acide carbonique combiné et l'oxyde de fer, la proportion de chacun de ces principes dans l'eau de mer elle même, et non dans le résidu de son évaporation: par ce moyen, chaque partie de l'analyse était aussi simple que peut l'être une opération de ce genre. (294, p.94)

He emphasized the analysis of sea water directly in its normal unconcentrated form and by the precipitative method. What is important about this work was that the problems of evaporation techniques of sea water analyses both with regard to total salts or specific components became painfully obvious. Prior to Usiglio virtually all of the sea water analyses relied on the evaporation of the sea water sample at least to some extent.

In comparing the representative analyses of sea water given above one easily notices that there is a wide latitude of data. With the detection of all of the major elements in sea water along with the many advances in chemical method and analytical techniques the precipitation method proposed by Murray should have been sufficient to produce consistent results, when used. There had been a tendency for many years in performing analyses of sea water to evaporate the sample to dryness and then often to redissolve the solid residue, usually in a smaller volume of water than that which had been evaporated, then the precipitative tests were performed if the analytical procedure was precipitative, or other solvents used on the dried residue if extraction was the technique used. By 1850 the precipitative method to determine the salts in sea water quantitatively was in general use. But this technique as it was practiced involved almost always the prior evaporation of the water. At the very least the sea water sample was evaporated to some extent so as to concentrate it. This practice was very common. The idea behind concentrating the sample was the belief that the precipitating agents would function more effectively in the more concentrated solution. Murray himself had done this as a matter of course in his analyses (226, p.212). This

TABLE XXVI

The analysis by Usiglio of three sea water samples partially evaporated to different specific gravities

Composition des sels	*Eléments*	*Poids obtenus pour 100 gr. d'eau de mer à 25 degrées*		*Poids pour 1 litre d'eau*	*L'analyse de l'eau de mer conduit aux resultats suivants*	
Analyse de l'eau à 25 degrées; densité 1,21						
Sulfate de chaux	ac. sulfuriq. chaux	0,1007 0,0705	0,1712	2,072	sulfate de chaux	0,165
Sulfate de magnésie	ac. sulfuriq. magnésie	1,2354 0,6360	1,8714	22,644	sulfate de magnésie	1,874
Chlorure de magnésium	chlore magnésium	1,8010 0,6410	2,4420	29,548	chlorure de magnésium	3,436
Chlorure de potassium	chlore potassium	0,1910 0,2130	0,4050	4,900	chlorure de potassium	0,402
Bromure de sodium	brome sodium	0,3360 0,0960	0,4320	5,227	bromure de sodium	0,428
Chlorure de sodium	chlore sodium	13,4870 8,7360	22,2230	268,898	chlorure de sodium	22,209
Poids total (gram)			27,5446	333,289	Total	27,504

(295,p.181)

TABLE XXVI (continued)

Une analyse semblable à la précédente a donné pour les eaux à 30 degrées les résultats suivants:

Indication des sels	*Eléments*	*Poids obtenus pour 100 grammes d'eau*		*Poids contenus dans 1 litre d'eau*	*100 grammes d'eau fournissent:*	
	Analyse de l'eau à 30 degrés; densité 1,264					
Sulfate	ac. sulfuriq.	4,110				
de magnésie	magnésie	2,121	6,231	78,76	ac. sulfuriq.	4,110
Chlorure de	chlore	5,930				
magnésium	magnésium	2,111	8,041	101,60	magnésie	5,560
Chlorure de	chlore	0,688				
potassium	potassium	0,761	1,449	18,32	potasse	0,916
bromure	brome	0,901				
de sodium	sodium	0,260	1,161	14,72	brome	0,901
Chlorure	chlore	10,220				
de sodium	sodium	6,610	16,830	212,80	chlore	16,838
Totaux (gramme)			33,712	426,20	on en déduit soude totale	8,56
						(295, p.182)
	Analyse de l'eau à 35 degrés; densité 1,32					
					moyenne de deux exper.	
Sulfate	ac. sulfuriq.	5,725				
de magnésie	magnésie	2,951	8,676	114,48	ac. sulfuriq.	5,7255
Chlorure de	chlore	10,913				
magnésium	magnésium	3,883	14,796	195,31	chlore	19,4475
Chlorure de	chlore	1,187				
potassium	potassium	1,310	2,497	32,96	brome	1,2005
Bromure	brome	1,2005				
de sodium	sodium	0,345	1,545	20,39	magnésie	9,2970
Chlorure	chlore	7,347				
de sodium	sodium	4,758	12,105	159,79	potasse	1,5790
Totaux (gramme)			39,619	522,93	on en déduit soude totale	6,881

(295, p.184)

was a direct carry over from the analysis of mineral waters, where the solutions in question were usually much more dilute than sea water and concentration was not only useful but almost necessary.[142] Surely the precipitation method and subsequent reporting of constituents was capable

TABLE XXVII

The analysis by Usiglio of sea water samples partially evaporated to varying specific gravities

Degrées de l'aréo-mètre de Baumé	*Volumes restant après l'évaporation et le dépôt*	*Dépôts obtenus à différentes densités*							
		Oxyde de fer	*Carbonate de chaux*	*Sulfate de chaux hydraté*	*Chlorure de sodium*	*Sulfate de magnésie*	*Chlorure de magnésium*	*Bromure de sodium*	*Chlorure de potassium*
3°5	1,000								
7 1	0,533	0,0030	0,0642						
11 5	0,316	–	traces						
14 0	0,245	–	traces						
16 75	0,190	–	0,0530	0,5600					
20 60	0,1445	–	–	0,5620					
22 00	0,131	–	–	0,1840					
25 00	0,112	–	–	0,1600					
26 25	0,095	–	–	0,0508	3,2614	0,0040	0,0078		
27 00	0,064	–	–	0,1476	9,6500	0,0130	0,0356		
28 50	0,039	–	–	0,0700	7,8960	0,0262	0,0434	0,0728	
30 20	0,0302	–	–	0,0144	2,6240	0,0174	0,0150	0,0358	
32 40	0,023	–	–	–	2,2720	0,0254	0,0240	0,0518	
35 00	0,0162	–	–	–	1,4040	0,5382	0,0274	0,0620	
Total des sels déposés		0,0030	0,1172	1,7488	27,1074	0,6242	0,1532	0,2224 (gram)	
Ajoutant les sels contenus dans les 0,0162 d'eau à 35 degrés					2,5885	1,8545	3,1640	0,3300	0,5339
Total des sels contenus dans 1 litre d'eau de mer		0,0030	0,1172	1,7488	29,6959	2,4787	3,3172	0,5524	0,5339
Analyse directe de l'eau de mer		0,0030	0,1170	1,760	30,1830	2,5410	3,302	0,570	0,518
Différences				–0,011	–0,4871	–0,0623	+0,0152	–0,0176	+0,0159

(295, p.185)

of a great deal of accuracy. But even though they were definitely better than those 40 years previous, by 1865 the analyses of sea water still showed appreciable variances.

Usiglio's work was conclusive proof that direct analysis was vastly superior in that it was capable of results that were reproducible and therefore supposedly more in keeping with the actual nature of the sea water analyzed. After the work of Usiglio there was a slow but definite trend away from evaporation in the analysis of sea water. For the next 20 years the analyses of sea water samples were not only much more consistent in themselves but comparable to one another.

After the work of Usiglio, analysis of mineral water ceased to be important to that of sea water. The two had been closely related for many years and indeed sea water analysis had risen out of mineral water analysis. The break between these two analyses was not a sharp one occurring with Usiglio but was one that had been in process since the early 1800's when men such as Marcet studied the make-up of sea water for its own sake, and not as a particular concentrated mineral water. With the precipitation method largely proposed by Murray and the demonstration by Usiglio that evaporation was not a quantitatively feasible procedure in sea water, the analysis of sea water went its own separate way accelerated by the new interest in the ocean that was beginning to take place at that time.

CHAPTER 6

CONSTANT PROPORTIONALITY OF CONSTITUENTS

It might seem, in theory, that a simple determination of total salt content by careful evaporation would yield a value that would be independent of the specific constituents. While this might be possible today, such definitely was not the case prior to 1865.

Sea water is essentially a solution of 11 major primary inorganic constituents which comprise almost all of the weight of the solid matter. These constituents or ions exist in sea water mainly uncomplexed.

There are a variety of difficulties, however, that arise in the determination of both the total salt content and the individual constituents. Upon evaporation of sea water to dryness the result is a complex mixture of inorganic salts. This mixture in itself may vary in composition depending on the manner of evaporation (rate, degree of heat, etc.). As the residue is heated, say to constant weight, water is driven off. The amount of water loss is a function of time and temperature. A number of the salts formed by the initial evaporation of the sea water contain water of hydration which varies in amount depending on the temperature of the drying process. Some of the constituent salts decompose in this heating. Magnesium chloride, for example, when heated gives off hydrogen chloride, leaving a basic salt of undetermined composition. At higher drying temperatures, some of the organic matter undergoes evaporation while some of it undergoes decomposition and oxidation. Elemental iodine may also be given off in the heating. Or simply, the weight of the salt residue will vary depending on the specific method used to produce it.

The determination of the individual components is also fraught with difficulties. There is, for example, still no satisfactory method to determine the concentration of sodium. The determination of sulfate by a barium salt is still easily the best method for this ion, yet this procedure involves much care in the control of precipitation conditions since co-precipitation of other ions with the barium sulfate is definitely a problem. The calcium precipitation, as the oxalate, also brings down the strontium, which involves a correction for this metal. Sea salt contains, then, primarily six constituents by weight. These are shown in Table XXVIII. And fluorine at 0.001 is sometimes also included.

The six constituents above represent the bulk (99.29%) of the total salt

TABLE XXVIII

The composition of sea water (1965)

Constituent	*G/kg of water of salinity 35‰*
Chloride	19.353
Sodium	10.76
Sulphate	2.712
Magnesium	1.294
Calcium	0.413
Potassium	0.387
These next four are also often considered major components:	
Bicarbonate	0.142
Bromide	0.067
Strontium	0.008
Boron	0.004

(63, p.122)

content in sea water. The second list of constituents (Table XXVIII) still contributes a fair amount of mass to the total (about 0.7%). The other minor constituents represent only 0.02–0.03% and may be considered negligible from a standpoint of salt content. It is obvious that prior to 1865 one must expect some definite variations in the determination of the salt content since almost 2% of the solids now known to be present had not then been accounted for. This is hardly a criticism. It would have been impossible for chemists to consider in their analyses elements that were not yet known to exist. Furthermore, it was natural to believe that substances known to exist per se should exist in the same form in solution. No one had any reason to conceive of a salt existing somehow separated in solution. It was known that chlorine (Cl^- or HCl) could be precipitated out of a salt solution by silver nitrate. A number of chemists felt, like Murray, that the "contents" of a solution should be expressed in terms of the "acids and bases", which together formed salts. Surely the elements themselves that formed sodium chloride could not exist separated in solution.

If one were to take a sea water sample and concentrate it slowly by evaporation, at some point in the process the salts would begin to separate out of solution, the least soluble first.[143] However, the results obtained from evaporation vary. This variance helps to point out the complexity of the problem. No one now would attempt to separate these salts by concentrating the water and collecting the fractions. These fractions have no sharply defined limits; each is contaminated with others. This would also be the case with the large scale salt production by evaporation. The salts produced were never pure. The insoluble matter would be largely calcium carbonate (along with some residual organic matter) primarily from the HCO_3^-, which becomes CO_3^{-2} upon heating. Very likely the carnalite ($KMgCl_3 \cdot 6H_2O$) existed primarily in the magnesium chloride.

The problem of sea water analysis was simply far more difficult than earlier chemists had thought. Because of this even a consistent description of the sea's saltness had not been arrived at. Boyle had never demonstrated that sea water is more salty one place than another. He had attributed the sea's saltness to be build-up of salt that rivers emptied into it. While Halley studied the sea in greater detail he did not cast much more additional light on these two points. To Halley the salt came constantly from the rivers and the salt in the ocean was constant from one place to another.[144]

In the years between Halley and Lavoisier there exists very little in the literature that takes issue with the origin of the sea's saltness or the variation of saltness from place to place. In the few analyses of sea water published in this period, these points were not mentioned (306, 309). The water samples chosen by these analysts were usually only from one locale. There was the implication that since the water does not vary from one place to another then it made little difference whether one chose water from one or several places.

When Lavoisier analyzed sea water he made reference to the origin of the salt in the sea (174, p.560) (see p.44 above). Bergman never mentioned the subject. Apparently he, too, believed that the sea received its salt from the world's rivers. This belief on the part of Bergman would not be unexpected. Bergman was very familiar with the geological thinking of his day (318, vol. 2, p. 420). The importance of the role the ocean played in the shaping of the Earth was emphasized by the schools of thought of probably the two most prominent geologists in the last part of the eighteenth century, the German Abraham G. Werner (1749–1817) and the Englishman James Hutton (1726–1794). Werner believed the sea was the most important factor in the origination of the various types of rocks. All the primitive rocks, he thought, were held in solution by the ocean and were chemically precipitated out (308, p.57). Hutton, on the other hand, believed the sea to be an important factor but certainly not more important than factors such as heat and air action (318, vol. 7, p.407). Hutton emphasized the role of running water as an erosive agent and carrier of substances such as salts to the sea (146, pp.iii, 18–19). Since the origin of the salt by river run-off was so reasonable and since personages like the great Lavoisier had said this to be his belief in one of the very few sea water analyses of late eighteenth century, then presumably there was little question about the subject. Lavoisier, like Bergman, published one analysis of sea water, although he had several samples. As only one analysis was published there is the implication that since sea water was of constant saltness that only one such analysis need be performed.

The influence of Bergman in water analysis was not only geographically far reaching but also extensive in effect. Since he became recognized as an authority of the analysis of all waters including sea water even though he published only one analysis of sea water, it is possible that this sole analysis of sea water had even more effect on the study of sea water than is readily apparent.

GAY-LUSSAC

Originally, the great French chemist Joseph Louis Gay-Lussac (1778–1850) had planned to analyze sea water to determine the nature and proportions of saline materials present (104, p.426). This idea was discarded once he became aware (104, p.427) of John Murray's analysis of water from the Firth of Forth. He felt that Murray's work had been done with great care and he saw no need to reproduce it. Rather Gay-Lussac decided to estimate only the total amount of saline matter present in sea water and to determine density as well.

Virtually all of Gay-Lussac's comments on the sea are contained in two articles, the latter being an addendum to the first, both published in 1817 (104, 105). The first, "Note sur la Salure de l'Océan Atlantique," (1817) is somewhat deceiving. In addition to dealing with salt content in detail, it also treated density at length as well as a number of related parameters and processes in the ocean such as temperature, fresh water influence, currents, evaporation, and precipitation. The paper was hardly a note but rather represents perhaps the first major treatise on the salt content of the oceans in general as well as several oceanic processes, both in terms of authority by a prominent scientist and coverage of content in an analytical and theoretical manner. This paper by Gay-Lussac represents one of the very best early attempts to explain a number of general phenomena of the sea and the first definite descriptive work on salt content.[145]

Presumably related to his studies of the composition of the atmosphere, Gay-Lussac became interested in the composition of sea water, especially with regard to density and salt content. Gay-Lussac had, in conjunction with the German scientist F.H. Alexander von Humboldt (1769–1859), shown that the composition of the air at elevation (about 7,000 meters) was the same as that of the surface.[146] His interest in the gaseous fluid, the atmosphere, seemed to stimulate his interest in the other great fluid system, the ocean. His attention seemed to be further whetted by the discordant and conflicting published accounts of the composition of sea water.

A number of sea water samples (see Table XXIX) were gathered from the middle of the English Channel, between Calais and Dover (104, p.428), by

TABLE XXIX

Density and salt content values by Gay-Lussac for a series of locations between Rio de Janeiro and France

Latitude	*Longitude*	*Densité*	*Résidu salin*
Calais	–	1,0278	3,48
35° (nord)	17° (ouest)	1,0290	3,67
31°, 50′	23°, 53′	1,0294	3,65
29°, 4′	25°, 1′	–	3,66
21°, 0′	28°, 25′	1,0288	3,75
9°, 59′	19°, 50′	1,0272	3,48
6°, 0′	19°, 55′	1,0278	3,77
3°, 2′	21°, 20′	1,0275	3,57
0°, 0	23°, 0′	1,0283	3,67
5°, 2′ (sud)	22°, 36′	1,0289	3,68
8°, 1′	5°, 16′	1,0286	3,70
12°, 59′	26°, 56′	1,0294	3,76
15°, 3′	24°, 8′	1,0284	3,57
17°, 1′	28°, 4′	1,0291	3,71
20°, 21′	37°, 5′	1,0297	3,75
23°, 55′	43°, 4′	1,0293	3,61
Moyenne		1,0286	3,65 (gramme)

(104, p.428)[152]

Gay-Lussac himself. Most of the samples were, however, supplied by an officer of the French Navy, in his return voyage from Rio de Janeiro (104, p.420). These samples, 15 in all, were taken at the surface and stored in tightly corked and tarred glass bottles.

Apparently only a month or so elapsed between the time of collection and before Gay-Lussac began analysis.

The densities were determined by weighing a vial, empty, full of distilled water, and full of sea water at a constant 8°C(104, p.427).[147]

Gay-Lussac agreed with Murray that the total salt content of sea water could be determined by an analysis such as Murray's simply by the addition of the weights of the individual components; but for determination of only the absolute salt content, he felt that a simple evaporation would work:

> ... mais il est plus simple et plus exact de la déterminer par l'évaporation poussée jusqu'au rouge obscur. Cette operation se fait très-commodement dans un matras dont on tient le col incliné sous un angle d'environ 45°, et que l'on agite continuellement pendant qu'il est sur le feu, aussitôt que les sels commencent à se déposer, afin d'éviter les soubresauts. Le bouillonnement ne peut rien projeter au déhors, et le résidu donné fait exactement le poids des matières salines.
>
> (104, p.427)

With the "bumping" solved by agitation and with the use of the long, thin, single neck of the matrass, the loss of material was negated – with one possible exception. It was not clear whether *l'acide hydrochlorique* (hydrochloric acid) came from the decomposition of a part of the *l'hydro-chlorate de magnésie* ($MgCl_2$) contained in sea water (104, p.427). This was allowed for by evaluating the

> quantité de cet acide en recueillant la magnésie qui reste lorsqu'on dissout le résidu de l'évaporation dans l'eau; car on connait le rapport d'après lequel ces deux corps se combinent. La quantité de magnésie fournie par chaque résidu étant trop petite pour être évaluée avec précision, on a réuni tous les résidus, et après en avoir séparé la magnésie, on l'a partagée proportionnellement au poids de chaque résidu. Comme il est très-probable que cette base existe dans l'eau de la mer à l'état de chlorure de magnésium, on a corrigé le poids de chaque résidu en eu retranchant celui de l'oxigène contenu dans la magnésie obtenue, et en ajoutant le poids du chlore saturé par la quantité de magnésium correspondante. (104, pp.427–428)

Gay-Lussac knew, then, that some hydrochloric acid might be given off during evaporation.[148] This, he believed, came from the magnesium chloride and that the residual product of the decomposition in the salt mixture was magnesia.[149] The oxygen content of magnesia would simply be mathematically converted to chlorine.[150]

The densities and salt contents varied in an irregular manner and did not necessarily correspond with one another. Here Gay-Lussac gave another possible reason for the variation in salinity.[151] He thought that it was possible that the disagreement in values for the residues was due to their not being heated to the same degree (104, p.428).

In consulting the work of previous analysts, Gay-Lussac decided, as Murray had before him (227, p.275), that these analyses were too inconsistent to trust (104, p.430). He concluded that the varying methods used must be the cause. In looking at Murray's total values for salinity (Murray did not use this term), he decided that these were too low but felt it was because the *salure* of this gulf is modified by the rivers which flow into it (104, p.430).[153] On the basis of his own values and those of others, he concluded that everything indicated that sea water contained at least *trois centièmes et demi* of salt matter. This value agreed with those of others. "Ce resultat s'accorde aussi avec la moyenne des degrées extrèmes de salure recueillis par mon ami M. de Humboldt" (104, p.430).[154]

It was difficult for Gay-Lussac to decide whether the salinity varied with latitude (104, p.429). He did believe that density and salinity were related. Although his results hardly showed any pattern, he did reach some conclusions.

> S'il était permis de tirer quelque conclusion de ces expériences, les densities prouveraient qu'à la latitude de Calais et de 10 N., la salure de l'eau est presque à son *minimum*; qu'elle est plus forte aux 35° et 32°N. de latitude; qu'elle va ensuite en diminuant jusque près de l'équateur, et qu'elle augmente dans l'autre hémisphère où, à la latitude de 17° à 24°, elle est la même que celle qui a lieu aux 35° et 32°N.
>
> D'après les résidus salins, la salure est à son *minimum* à la latitude de Calais et à celle de 10° N. Elle augmente ensuite, quoique d'une manière irrégulière, et paraît un peu plus forte dans l'hémisphère austral que dans l'hémisphère boreal. Ainsi, en comparant les densités aux résidues salins, il semblerait que la salure de la mer est moindre dans le canal de la Manche, et à 10° de latitude N., que par-tout ailleurs, et qu'elle est un peu plus grande pour l'hémisphère austral que pour le boréal. (104, p.429)

After much consideration, Gay-Lussac decided that the *salure* of the great ocean has very small variations, if it is not the same everywhere (106, p.432).

This is extremely important. This is *the first precise pronouncement that the salinity of the open ocean (specifically the Atlantic) is constant.*

Considering the inconsistency of some of his data, it is surprising that he should arrive at this pronouncement. If, however, one considers that earlier he had shown that at least to the altitude of 22,000 ft. the earth's other fluid ocean, the air, was of constant composition, then it is less surprising. Based on his work with air, Gay-Lussac probably believed when he began his tests that the ocean was equally salty from place to place. One reason for the equality of saltness was the constant mixing of the oceans. "Mais les courans continuels que l'on sait exister dans les mers tendent toujours à rétablir l'équilibre de salure" (104, p.433). It should be remembered that these comments dealt specifically with the surface. His water samples had not been taken at depth.

Any variation in the salinity of the ocean was, to Gay-Lussac, a local phenomenon caused by some local condition (104, p.432). The deviation toward a lower salinity would, he explained, be due to river run-off and precipitation. Higher values would be due to greater evaporation than precipitation as might occur in regions such as the equator. Gay-Lussac (106, p.433) thought that fresh water and sea water do not mix rapidly, but that the fresh water, as from a river, might be found well away from land.[155]

From the small amount of data available from the surface samples and with some ideas of atmosphere homogeneity from Humboldt (144, 145). Gay-Lussac stated that "On peut par consequent (an extrapolation of atmospheric work) regarder comme très-probable que la salure des mers est sensiblement la même au fond et à leur surface" (104, p.435).[156] This inference that the salt content was the same from surface to bottom was based on several things. Gay-Lussac knew that sea water was definitely not saturated with saline matter and thus he felt temperature had little to do with salinity. Furthermore, he had no data that showed that temperature changed with depth. Finally, he felt that even if temperature did change with depth it would be a function of cold currents coming from the poles (104, p.436). In all, he felt that temperature, like salinity, would be averaged out in the ocean by currents. "Ainsi, les courans des mers, comme ceux de l'Océan aerien, tendent continuellement à établir un équilibre de température sur la surface de notre globe" (104, p.436). Note the analogy to the ocean of air.

This question of possible variations of salinity with depth continued to bother Gay-Lussac and later he became less secure than in his original statement. In the second paper, a supplement to the first, he said that: "Les expériences que je connaissais sur la salure de la mer à de grandes profondeurs ne me paraîssaient pas mériter une grande confiance" (105, p.80). Indeed, this second of his two papers on the salinity of the sea (105) dealt specifically with this topic.[157]

In his first paper on sea water, Gay-Lussac had mentioned a water column, "la salure d'une colunne d'eau de l'Océan" (104, p.435). This was new.[158] In attempting to study this phenomenon, he stored in a vertical position for up to 20 months tubes of glass, with dimensions of 2 cm by 2 m, filled with saturated solutions (105, p.82). Later he also tried unsaturated solutions of varying salts (106, p.306). All of these tubes were stored at a constant temperature of 11.67°C (106, p.299) in caves for up to 20 months. The amount of salt at the top and bottom was ascertained by the aforementioned evaporation in a matrass.[159]

Although he had some reason to believe that salt water should increase in saltness with depth,[160] Gay-Lussac was never able to demonstrate this. His research on the tubes gave equal amounts of salt from top to bottom. As a consequence, he did not change his earlier stand on salinity with depth.[161] These tests did, however, enable Gay-Lussac to conclude one very important thing: specifically, that there was no separation of substances dissolved in water due to differences in density. "D'après ces divers résultats, je regarde comme certain que les molécules salines d'une dissolution saturée, dont la température est supposée constante, ne s'en séparent point en vertu de leur plus grande densité" (107, pp.82–83).

These two articles represent Gay-Lussac's direct contribution to the study of the sea. But perhaps an even greater contribution came in a more indirect way. He was intensely interested in virtually all phases of science. After his papers on salinity, he became increasingly more interested in analytical chemistry, especially in the volumetric aspects. Gay-Lussac was one of the greatest of all analytical chemists. Most analytical chemists had been interested in mineral analysis but primarily from a gravimetric standpoint. Gay-Lussac did not analyze minerals – a departure from the then accepted work of chemists. But he was familiar with gravimetric methods, and indeed, he used them often. Between the years 1824 and 1835, Gay-Lussac published a number of papers on volumetric analysis. Although these papers on this subject continued almost until his death, the bulk appeared during this interval.[162]

In a broader sense, the works of Gay-Lussac in titrimetry and volumetric analysis gave chemistry rapid, simple and accurate methods in analysis that were increasingly used in chemistry and in the study of the sea.

In addition to his chemical work (see p.69) Alexander Marcet made a number of general comments about salinity and its relationship to density (199). He characterized the density as being a measure of the saline contents of the water.[163] The higher the density the higher the salt content.[164] From these densities Marcet concluded that:

> The ocean in the southern hemisphere, would appear to contain more salt than in the northern hemisphere, in the proportion of 1029, 19 to 1027, 57; as may be seen by taking the mean specific gravity of the waters collected from the two hemispheres. But it must be observed, that a great proportion of the specimens from the northern hemisphere were taken farther from the equator than those procured from the other hemisphere, which may possibly account for the difference in question.
>
> The mean specific gravity of specimens taken from various parts of the equator is 1027, 77, and is therefore a little greater than that which prevails in the northern hemisphere, though sensible less than that of the southern ocean.
>
> There is no notable difference between different east and west longitudes at the equator; nor is there, in other latitudes, any material and constant difference between waters of the ocean in corresponding east or west longitudes in the same hemisphere.
>
> There is no satisfactory evidence, at great depths, being more strongly impregnated with salt than it is near the surface. In general the waters of the ocean, whether taken from the bottom or from the surface, appear to contain most salt in places in which the sea is deepest or most remote from land; and the vicinity of large masses of ice seems to have a similar effect to that of land in diminishing the saltness of the sea. It may be stated generally, that small inland seas, though communicating with the ocean are much less salt than the open ocean. This is particularly striking in the case of the Baltic; and also, though in a less remarkable degree, in the Black Sea, in the White Sea, in the Sea of Marmora, and even in the Yellow Sea.

> The Mediterranean, though a comparatively small and subordinate sea, is found to contain rather a larger proportion of salt than the ocean.[165] (201, pp.173–174)

Marcet said that salinity decreased with depth but that the rate of decrease seemed to vary from place to place (201, p.183). With regard to temperature, he remarked that this data:

> uniformly led me to the conclusion, that the law of greatest specific density at 40° [F], did not prevail in the case of sea water; but that, on the contrary, sea water increased in weight down to the freezing point, until it actually congealed.[166] (201, p.187)

He concluded from this work on sea water:

> With the exception of the Dead Sea, and of the Lake Ourmia, which are mere salt ponds, perfectly unconnected with the ocean, all the specimens of sea water which I have examined, however different in their strength, contain the same ingredients all over the world, these bearing very nearly the same proportions to each other; so that they differ only as to the total amount of their saline contents. (201, p.194)

This statement of Marcet's is probably *the first suggestion of the relative constancy of composition of sea water*. While he knew that the total salt content could vary from place to place, he felt that the variation in proportional composition of constituents was at best trifling. In fact, he seems to have been of this opinion before he began the analyis:

> ... however unlikely to be productive of any striking discovery, such an inquiry, conducted with due care and attention, might afford curious results, and throw some light on this obscure subject (composition of sea water) (201, p.161)

The words "however unlikely" imply that he believed that either the total salt content of the open sea (away from other influences) or the relative components or both to be a constant. It is suggested that he believed probably both but certainly the latter to be constant.

There was nothing in the literature of the times that would indicate a general belief in constancy of proportions of salts in sea water per se. But there seemed to be at this time the prevalent idea as to constancy in the "world's fluids". Gay-Lussac had said the air was constant in composition and that the sea was constant in salinity. Davy, as a result of his density experiments, said in 1824 that "the results favour the general conclusion already formed by some philosophers, that the ocean resembles the atmosphere in being ("caeteris paribus") of nearly the same specific gravity throughout" (69, p.319).

Furthermore, Davy added: "That the specific gravity of the water of the ocean, in all its parts, however remote, should be nearly the same, is easily explained; it is indeed what might be expected from theory" (69, p.320). The theory he referred to is similar to Gay-Lussac's idea of mixing by circulation causing equally distributed and even saltness. Dr. Alexander Fyfe (see p.77), while his results were erratic, concluded that "The results of the above experiments show, that the water of the ocean, from north latitude 61° 52′ to north latitude 78° 35′ does not differ essentially in the quantity of saline matter which it contains" (101, p.162). The idea that sea water was of constant saltness probably was further driven home to Marcet by the only slight variation in the densities he determined. It does not seem unreasonable to suspect that his idea of constant proportional constituent composition was a direct outgrowth of this.

The question of constancy of salinity was, however, hardly settled, and articles appeared which took issue with this supposed constancy. With regard to the variation in salinity (*salure*) Darondeau (see p.77) said:

> Si l'on considère la proportion des résidus provenant de la proportion des résidus provenant de la dessiccation, on voit, comme dans le cas précédent, que généralement l'eau de mer a un degré de salure plus considérable au found qu'à la surface; dans un cas, cependant, le degré de salure est moindre. (66, p.104)

Darondeau used for his "salure" determination exactly the method of Gay-Lussac (see earlier). There was no attempt to compare surface salinities in general – only surface versus depth per locale. The data, Darondeau realized, was hardly conclusive, but he felt it was adequate to generalize that the surface salinity was less than that at depth.

By 1850 it was becoming not uncommon to refer to the idea of constancy of constituents. Dr. Friedemann Göbel (114, 115) who, as mentioned above (p.75), made a quantitative determination of bromine in sea water, had this to say concerning the constancy of proportions of constituents:

> Die Bemühungen wissenschaftlicher Reisender, besonders aber von Gay-Lussac, Marcet, Tennant, Davy, Wollaston, Horner und in den neuesten Zeiten von Lenz ausgeführte Arbeiten über die spezifischen Gewichte des Wassers der verschiedensten Meere haben dargethan, dass das Meer in seinen verschiedenen Abtheilungen ein nur wenig von einander abweichendes Bestandtheilverhältniss hat. Ueberall finden wir dieselben Salze, und nur in quantitativer Hinsicht zeigen sich in verschiedenen Meeresabtheilungen einige, jedoch äusserst geringe, Differenzen. (115, p.104)

Not only did Göbel know that the salt content in the ocean varied from place to place, but also there is little question that Göbel believed in the constancy of the ratio of "acids and bases."

The values for the salt content of surface waters of Ernst von Bibra (see p.75, above) varied substantially from one another. While he explained the variances in the offshore values (16, p.98), he neatly neglected any mention of the variances in the open oceans. A possible reason for this was his difficulty in reconciling the data with his belief which he probably harbored that the salt content was a constant with respect to geographical location but not with depth. There was no question that he believed the salt content of the water near the bottom was greater than that of the surface.

> Quantitative Verschiedenheiten finden indessen ziemlich bedeutender statt, wie solches schon aus der Summe der festen Bestandtheile überhaupt hervorgeht. Es werden Dieselben ohne Zweifel einerseits sowohl durch die Verschiedenheit des Meeresbodens selbst bedingt, als andererseits auch Strömungen und vielleicht selbst durch Stürme, wenigstens was die Oberfläche der See betrifft. (16, p.98)

Von Bibra seemed to be of the opinion, though not strongly so, that the constituents in sea water were identical whether at the surface or at depth. He also seemed to believe that the constituents of the seas were the same from place to place and that there was a proportionality between them. This was almost part and parcel of the belief that the ocean was, except in certain locales, generally of equal saltness.

> Aus diesen Versuchen scheint hervorzugehen, dass die qualitative Zusammensetzung des Meerwassers, wenigstens die des von mir untersuchten und an den bezeichneten Stellen geschöpften, dieselbe ist, insoferne nämlich bloss die angegebenen Bestandtheile im Auge behalten werden. (16, p.98)

However, some chemists believed in neither the constancy of saltness nor the constancy of constituents. August A. Hayes, Assayer to the Commonwealth of Massachusetts and a prominent American chemist, published in 1851 an article dealing with chemical conditions in sea water (132). Although the paper was not concerned with salt content but was instead a detailed discussion of corrosion in the sea, Hayes mentioned that the prevailing belief at that time was that the sea had existed for great periods of time with the same amount of saline matter dissolved in it. He knew that local deviations such as evaporation or decomposition of rocks disturbed this apparent uniformity and believed that as the topic was studied more especially with respect to the decomposition product of rocks, that this idea would gradually give way. Hayes doubted this constancy of constituents.

During the first half of the nineteenth century there was more of a trend toward the study of the ocean in general than the scattered chemical analyses would indicate, although these studies, too, were rather sporadic. The most consistent study of the sea during these times was in the realm of marine zoology. The observations that Captain Robert Fitz-Roy and Charles Darwin made between 1831 and 1836 on the voyages of the "Beagle" and the subsequent book *The Voyage of the "Beagle"* (67) did much to stimulate interest in marine biology. The two main results of Darwin's voyage, namely his theory of coral atoll and reef formation (especially the former) and his theory of evolution, gave rise to and was an impetus in a number of scientific questions concerning oceanography.

There were two men in this time period that exhibited more than the usual passing interest in the sea. They were the Frenchman Georges Aimé and the American Matthew Fontayne Maury (1806–1873).

There is little question that Aimé was an oceanographer. As his eight collected memoirs (2) indicate, he was primarily interested in the physical aspects of the sea, which is understandable in view of the fact that he was a professor of physics at the College of Algiers (240, vol.4, p.912). He spent most of his waking hours, and he slept little, in the study of the sea. The list of devices he developed is impressive.[167] Unfortunately, few Frenchmen followed in his footsteps. Perhaps one of the reasons was his short life – he died at 35 due to a fall from a horse. When Louis Thoulet set up his marine center at Nancy at the turn of the century and subsequently wrote his book on oceanography (289), he was unfamiliar with Aimé's work.

MAURY

If one were to look at any of the pilot charts issued by the Hydrographic Office of the United States Navy Department, he would find at the top of all of these the words: "Founded upon the researches made and the data collected by Lieutenant M.F. Maury, U.S. Navy." In 1848, Maury published a pamphlet of only ten pages called *An Abstract Log for the Use of American Navigators* (179, pp.53–54). Future editions were issued as the *Sailing Directions* which eventually contained 1257 pages in two volumes in quarto.[168] As Maury issued the various editions of the "Sailing Directions" he added increasing amounts of descriptive matter concerning the sea. The sixth edition of "Sailing Directions" (1854) contained 90 pages of this descriptive matter. The title for his book, *The Physical Geography of the Seas*, was a chapter heading in this sixth edition.[169] In early 1855 the first edition of Maury's book, "The Physical Geography of the Seas," was published. This book went through eight editions in the United States, increasing from 274 to 474 pages; the last appeared in 1861.[170]

TABLE XXX

The salt contents of sea water given by Maury

Water	962.0 grains
Chloride of sodium	27.1
Chloride of magnesium	5.4
Chloride of potassium	0.4
Bromide of magnesia	0.1
Sulphate of magnesia	1.2
Sulphate of lime	0.8
Carbonate of lime	0.2
Leaving a residuum of	2.9
	1000

(211, p.17)

The first edition of "The Physical Geography of the Seas" (1855) was written hastily (151, p.106). It contained, probably for this reason, a large number of errors, most of which were in content. In later editions most of these were rectified. Reading "The Physical Geography" from a scientific standpoint, however, is difficult, especially with respect to the subject of the salt in the oceans. Maury generally mentioned some source references in treating of a topic. On the question of the salt content of sea water these references were lacking. The salt content for 1000 grains of sea water was given by Maury as is shown in Table XXX. No authority was cited for this.[171] Maury, however, did mention these solid ingredients: "Solid ingredients: iron, lime, silver, sulphur, and copper, silex [silica], soda, magnesia, potash, chlorine, iodine, bromine, ammonia, are all found in sea water" (212, p.16). This is reminiscent of the manner of expressing the constituents as "acids and bases" although copper, silver, and sulphur, as such, usually would not have been included with these others. He made no attempt to explain these terms–salt content, and solid ingredients–especially by any cross reference.

Since Maury mentioned Marcet more than anyone else (211, pp.i, 32, 184) with respect to the chemistry of the sea, it seemed reasonable to suppose that this data came from him. Yet the list of "constituents" is more recent than Marcet since it contains many more compounds which Marcet had not found in sea water. These results could have come virtually from anyone. But a check of all of the previous analyses does not reveal a single set of values that agrees exactly with Maury's. Maury was no chemist and he did not analyze sea water himself. The total salt content is high: 3.8% (salinity of 38‰). This is higher than his value given in the 1858 edition (6th) where he stated: "The solid constituents of sea water amount to about 3½ percent of its weight, or nearly half an ounce to the pound" (21, p.184). This edition (6th) did not enumerate the salts as in the earlier edition, but only stated: "They are chiefly common salt, sulphate and carbonate of lime, magnesia, soda, potash, and iron" (212, p.184). A footnote did add one more compound: "It is the chloride of magnesium which gives that damp, sticky feeling to the clothes of sailors that are washed or wetted with salt water" (212, p.184).

The high salt content of 3.8% suggests the Mediterranean as its source and Maury probably saw it in this book:

> But by far the most interesting and valuable book touching the physical geography of the Mediterranean is Admiral Smyth's last work, entitled "The Mediterranean; A Memoir, Physical, Historical, and Nautical," By Rear-admiral William Henry Smyth, K.S.F., D.C.L., and c. London; John W. Parker and Son, 1854.
>
> (212, p.XIV)

But the analyses in Maury's 1861 edition did not come from Smyth's work. Presumably the data came from authors such as Von Bibra. Evidently this data was not available to Maury in 1855 and was added when it came to his attention. Most likely, however, the analysis in the 1861 edition was an average done by Maury of the works of several analysts. The following quotes from Maury are significant. "As a general rule, the sea is nearly of a uniform degree of saltness, and the constituents of sea water are as constant in their proportions as are the components of the atmosphere" (212, p.16).[172] The constancy of the ocean's saltness and especially that of the specific salts is strongly asserted here in the 1861 edition. Yet a similar but even stronger statement had appeared earlier in the 1855 edition of "The Physical Geography of the Sea":

> The salts of the sea, as its solid ingredients may be called, can neither be precipitated on the bottom, nor taken up by the vapors, nor returned again by the rains to the land; and but for the presence in the sea of certain agents to which has been assigned the task of collecting these ingredients again, in the sea they would have to remain. There, accumulating in its waters, they would alter the quality of the brine, injure the health of its inhabitants, retard evaporation, change climates, and work endless mischief upon the fauna and the flora of both sea, earth, and air. But in the oceanic machinery all this is prevented by compensations the most beautiful and adjustments the most exquisite.
>
> (211, p.14)

Again it is difficult to say what authority he used for this first statement. The best answer would probably be Marcet. Both of these statements were, however, more strongly phrased than was Marcet's (see p.96). There is also no indication as to why Maury toned down this idea in the 1861 edition. But the fact remains he did seem less certain about it. It is possible that this statement might have come under fire, especially with regard to the idea of constant saltness. Maury, a religious man, believed firmly in the harmony of the world. It is suspected that he believed that the total salt content as well as the salt constituents themselves could not be other than constant.

Maury's declaration that the salt content and the constituents thereof were constant was important. Maury stated this unequivocally and others had done so prior to him. But Maury's comments had more effect on the general thinking of this time than these others. To understand this one has to consider the esteem in which Maury was held. Maury was unquestionably the first man to chart the world's currents and winds. By 1851 he had over 1000 ships of all nations at any given time sending him current, wind, and meteorological data (179, p.80). Travels on the sea after Maury's "Sailing Directions" were not only safer but faster (177, p.82). For example, before Maury, it took 180 days to sail from New York to San Francisco. His charts cut this to 130. With his charts the most famous of the clipper ships, the *Flying Cloud*[173], under Captain Josiah P. Cressy, made the run in 89 days. In short, Maury, owing to his immense contributions to sea travel, was one of the most famous men of his day. He was honored by most of he Crowned Heads of Europe and given numerous titles. His book, "Physical Geography of the Seas," was easily the first widely read book on the sea. This book had a great deal of popular appeal – a fact not generally known. It was the first book on the seas that had any impact on the public, scientific and non-scientific alike. In such a book by such an accepted authority it seems reasonable to suppose that material contained therein should be unquestionably acceptable by many readers. Besides, the idea of a homogeneous ocean was esthetically pleasing.

Largely due to the strength of his reputation, Maury, although not a chemist, was more responsible for the belief in the constancy of salt content and proportion of constituents of sea water than anyone else before him.

In perspective, one should mention that Maury's study of he sea probably could not have been nearly so fruitful if there had not existed at this time a widespread desire to learn more about the oceans. It was in this time period from about 1830 on that men began to study the sea. Expeditions literally in the course of their travels began to explore the depths of the sea. Maury's work and his book "Physical Geography of the Seas" aided in this study.

Before the modern concept of salinity-chlorinity could develop, several things had to come to pass. First, the general acceptance of the concept that the salinity is constant in the world's oceans. The work and suggestions of Murray gave chemists a method, which, when followed with care, was capable of giving reproducible results in sea water analysis. The word "salure" is an old French word meaning saltness. It had been used for years in reference to the deposits remaining after the solar evaporation of sea water for salt. In a modern sense, the word could take on the meaning of salinity. This is the way Gay-Lussac seemed to use the term. With Gay-Lussac there was the first use of a concept of salinity as it is used in the modern sense. Secondly, there had to exist the belief in the proportional constancy of constituents – this, too, resulting from analysis. The comments of Marcet were in large part the effective originator of this idea whereas Maury was largely responsible for its spread.

By 1865 the situation would appear to be this: the non-scientific world viewed the salt content of the sea as a constant. The bulk of the scientific community knew that the salt content varied at least in-shore and for seas such as the Mediterranean, but little else had been said. The data was not conclusive one way or the other for the open sea. There were no firm statements on the salinity of the world's oceans at large. But a number of researchers believed that definite surface variations did exist.

With regard to the relative proportions of constituents things appeared to be more settled. While the data was by no means conclusive, the general feeling prevailed, and increasingly from 1820, that these constituents were constant in proportion.

CHAPTER 7

THE COEFFICIENT

Between the years 1843 and 1865 the great Danish chemist Johann G. Forchhammer published several papers on the composition of sea water. The treatise published in 1865 in the Royal Society's *Philosophical Transactions* entitled "On the Composition of Sea-water in the Different Parts of the Ocean" (100) represents the culmination and essentially the sum total of his endeavors for over 20 years of patient work. It is primarily in this paper, rather than in earlier ones, that Forchhammer presented his conclusions on sea water. The 1865 (100) paper was, however, a larger, more detailed version of one of 1859 (97). The primary reason for the six year delay is simply the fact that with the extensive task of the analyses themselves plus his other research, especially his exhaustive duties at the Copenhagen Institute, he found little leisure time to interpret his results as he obtained them.

Johann Georg Forchhammer (1794–1865) received his Ph.D. in 1820 from Kiel University where he majored in chemistry and pharmacy (255, p.13). He became Professor of Mineralogy at the Polytechnic Institute at Copenhagen and succeeded the famous physicist, Hans Christian Oersted (1777–1851), as Director of this institution. For years he was Secretary of the Danish Royal Academy of Science.

Forchhammer was an extremely adept and industrious chemist. Between the years 1830 and 1865 he published an amazing number of papers on a wide range of chemical topics.[174] As for his interest in the chemistry of the sea, he explained that the receipt of a number of sea water samples from a friend in 1843 had led him toward sea water analyses.

> In the year 1843 a friend of mine, Mr. Ennis of Falmouth, sent me some bottles of sea-water from the Mediterranean, which I subjected to a chemical examination, a work which induced me to collect what other chemists had determined about the constitution on the water of the great Ocean. This labour convinced me that our knowledge of the composition of sea-water was very deficient, and that we knew very little about the differences in composition which occur in different parts of the sea.
>
> I entered into this labour more as a geologist than as a chemist, wishing principally to find facts which could serve as a basis for the explanation of those effects that have taken place at the formation of

> those voluminous beds which once were deposited at the bottom of the ocean. I thought that it was absolutely necessary to know with precision the composition of the water of the present ocean, in order to form an opinion about the action of that ocean from which the mountain limestone, the oolite and the chalk with its flint have been deposited, in the same way as it has been of the most material influence upon science to know the chemical actions of the present volcanos, in order to determine the causes which have acted in forming the older plutonic and many of the metamorphic rocks. Thus I determined to undertake a series of investigations upon the composition of the water of the ocean, and of its large inlets and bays, and ever since that time I have assiduously collected and analysed water from the different parts of the sea. (100, p.203)

Since his major professor Pfaff (see p.75) was, at the time Forchhammer was at Kiel, working on sea water analyses, it is not unreasonable to suppose that he was at least familiar with this well before the 1843 date above. But as Forchhammer said in 1845;

> For at kunne forklare alle Havets neptuniske Dannelser er en nöiagtig Analyse af Söevandet fra de forskjellige Dele af Verdenshavet uundgaaelig nödvendig, og Forfatteren har begyndt paa dette omfattende og möisommelige Arbeide.[175] (92, p.28)

Thus, it may have been that Forchhammer asked Mr. Ennis to send him sea water samples for his geological work. The reason was to explain deep water sediment formations. To do this he needed an exact analysis of sea water from around the world.

Forchhammer's paper was a milestone in the history of chemical theories of sea water. In discussing Forchhammer and his work the term salinity has been used often. The reason is simple. Forchhammer introduced this term.

> The next question to be considered refers to the proportion between all the salts together and the water; or to express it in one word, I may allow myself to call it the salinity of the sea-water, and in connection with this salinity or strength, the proportion of the different solid consituent parts among themselves. (100, p.219)

After Forchhammer, this term would become common in reference to the total salt content of sea water. In addition to new terminology Forchhammer's paper was an extensive definitive study of the composition of sea water that was well read, widely cited and influential. From the standpoint of salinity-chlorinity it was the most important paper written by 1865. In all Forchhammer determined 27 elements in sea water although not all were determined quantitatively (100, pp.205–214), and he detected several of these for the first time in sea water. He regarded as impractical the determination of all of these 27 elements in sea water analyses. Since many of these were in such small amounts the water volume needed to determine them would be in the realm of 100 lb., which would be too difficult to obtain and in itself, too prone to change due to evaporation and fermentation (100, p.214). In order to avoid this extremely difficult and time-consuming analysis, since he felt that such a laborious procedure would never produce the data needed for the comparison of the oceans' waters, Forchhammer determined chlorine, sulfuric acid, soda, potash, lime and magnesia. Occasionally he determined silica, phosphoric acid, carbonic acid and oxide of iron although they were present in rather small amounts.

> All the numerous other elements occur in so small a proportion that they have no influence whatever on the analytical determination of the salinity of sea-water, though, on account of the immense quantity of sea-water, they are by no means indifferent, when we consider the chemical changes of the surface of the earth which the ocean has occasioned, or is still producing. (100, p.214)

Forchhammer did not use the evaporation method:

> I rejected a method often used, which consists in evaporating sea-water to dryness, because it is inaccurate, and the result depends partly upon trifling circumstances. If evaporated by steam of 100°C. there will remain a very notable quantity of water, which quantity can only be ascertained with great difficulty. If it is dried at a higher temperature muriatic acid from the chloride of magnesium will be driven out together with the water. I preferred thus, as I have already mentioned, to determine the quantity of the five-above-named substances, to ascertain under one head all the small quantities of the different substances that remain insoluble in water, such as silica, phosphate of lime, etc. and to calculate the soda. (100, p.215)

His analyses were entirely gravimetric, his methodology being much like that of Murray with obvious improvements. An example is his determination of chlorine[176]:

> Of one portion of 1000 grains, I separated the chlorine by nitrate of oxide of silver after I had poured a few drops of nitric acid into the water. In those cases where the water had fermented, I allowed it to stand in an open glass jar, in a warm place, until all smell of sulphuretted hydrogen had disappeared. To try how exact a result this method could give, I took a larger portion of sea-water, and weighed three different portions, each of 3000 grains, and precipitated the chlorine. The result was: chloride of silver 145.451, 145.544, 145.642, mean: 145.541. The

greatest difference is: −0.090 = 0.022 chlorine, + 0.083 = 0.020 chlorine.

These small differences are probably due to the small irregularities occasioned by the evaporation of very small quantities of water during weighing. The dried chloride of silver was as much as possible removed from the filter, melted in a porcelain crucible, weighed, and calculated as pure chloride of silver. The filter was burnt in a platinum crucible, by which the chlorine which had been combined with it was calculated. This supposition is correct if the quantity of chloride of silver adhering to the filter is very small. (100, pp.215–216)

Forchhammer gave the following criteria for the combination of the acids and bases to produce salts.

The whole quantity of lime was supposed to be united with sulphuric acid. What remained of sulphuric acid after the saturation of lime, was supposed to be combined with magnesia. What remains of magnesia after the saturation of sulphuric acid, was supposed as magnesium to enter into combination with chlorine, and form chloride of magnesium. The potash was supposed to form chloride of potassium. That portion of chlorine which was not combined with magnesium or potassium, was supposed to form a neutral combination with sodium.

Lately, that small quantity of different substances, "silica etc."[177] was added, and the sum of all these combinations thus calculated forms the number which in the Table is called "All salts." It is hardly necessary to remark, that it is quite indifferent how we suppose the acids and bases to be combined in sea-water, the sum must always be the same, provided the salts are neutral, and all the acids (chlorine included) are determined, as well as all the bases, with the exception of soda.

(100, p.219)

His results are tabulated in Table XXXI. He included similar data for the North, Middle, and South Atlantic and Indian Oceans, the North, Black, Caspian, Red, Asov, and South Polar Seas, as well as the Baltic, Caribbean, Mediterranean (Table XXXII). Most of the Pacific Ocean was lacking.

In an effort to use the tremendous amount of data that he had gathered, Forchhammer included a number of comparison tables (see Appendix VI). He made every attempt at accuracy and was often handicapped by the length of time the samples, brought to him by seamen, had sat about before analysis. He tried to allow for this:

It might seem that the relative quantity of salt might be inexact, because water might have evaporated through the cork during the long time which often elapsed between the time when it was taken up from

TABLE XXXI

The composition of sea water in the different parts of the ocean between latitude N 51°1′ and 55°32′; and longitude 12°6′ and 15°59′ by Forchhammer

Magnesia	*Silica etc.*	*Chloride of sodium*	*Sulphate of magnesia*	*Sulphate of lime*	*Chloride of potassium*	*Chloride of magnesium*	*All salts*	*Coefficient*
2.211 (11.24)	0.110	27.977	2.376	1.353	0.700	3.212	35.728	1.816
2.211 (11.18)	0.100	28.056	2.279	1.483	0.603	3.344	35.865	1.814
2.235 (11.35)	0.074	27.735	2.213	1.402	0.686	3.438	35.548	1.805
2.226 (11.30)	0.105	28.005	2.373	1.385	0.581	3.305	35.754	1.814
2.179 (11.03)	0.071	28.119	2.298	1.409	0.685	3.206	35.788	1.812
2.175 (11.06)	0.071	27.914	2.193	1.487	0.575	3.330	35.570	1.809
2.128 (10.83)	0.071	28.139	2.279	1.418	0.531	3.145	35.583	1.811
2.209 (11.18)	0.078	28.188	2.451	1.369	0.517	3.203	35.806	1.812
2.145 (10.92)	0.113	28.119	2.355	1.354	0.592	3.131	35.664	1.815
2.183 (11.24)	0.104	27.740	2.432	1.359	0.555	3.158	35.348	1.820
2.225 (11.34)	0.088	27.916	2.379	1.326	0.517	3.298	35.524	1.811
2.182 (11.08)	0.069	28.081	2.253	1.457	0.511	3.261	35.632	1.810
2.192 (11.15)	0.090	27.983	2.320	1.377	0.581	3.263	35.615	1.811
2.193 (11.14)	0.086	28.011	2.326	1.417	0.502	3.245	35.677	1.813

(100, p.261)

TABLE XXXII

The composition of sea water from the Red Sea, and from different depths in the Baltic, by Forchhammer

Magnesia	*Silica etc.*	*Chloride of sodium*	*Sulphate of magnesia*	*Sulphate of lime*	*Chloride of potassium*	*Chloride of magnesium*	*All salts*	*Coefficient*
2.685 (11.31)	0.136	33.871	2.882	1.676	0.612	3.971	43.148	1.818
0.403 (12.38)	0.027	4.474	0.329	0.322	0.089	0.678	5.919	1.818
0.441 (11.14)	0.072	5.810	0.632	0.333	0.092	0.526	7.465	1.886 (grams)

(100, p.261)

> the sea, and the time when it was analyzed. It is, however, easy to see whether the quantity of water in the bottle has diminished, or whether the cork has been corroded; in both cases the sample has been rejected, but I must remark that these cases have been rare. In the last three or four years all the samples which have been taken according to my direction have been marked on the neck of the bottle with a file, on that place to which the water reached when the bottle was filled.
>
> (100, pp.218–219)

Forchhammer used many of these data in an attempt to trace water masses, such as the Gulf Stream (100, pp.220–221, 242) and he predicted the existence of a sub-surface Antarctic Current before it was actually discovered (100, p.226).[178] Although this paper contained no actual analyses of river water, Forchhammer evidently had run a number of them since he was interested in the difference in the proportion of the constituents in river water versus sea water (100, p.243). He attributed the smaller quantity of carbonate, lime and sulfate of lime which predominate (as bases) in river water, but not in sea water, to the action of shell fish which removed these substances. These variances between sea and river water had been noted for some time but Forchhammer quantified and explained them (100, p.245). He offered a similar explanation for the difference in silica content between river and sea waters.

The overall salinity of the open oceans Forchhammer regarded as *almost* constant with the exception of areas such as the North Sea and the Baltic (100, p.242). This constancy he felt was due largely to the tremendous mass and constant motion of the ocean (100, p.226). Local variations were to be expected near the continents due to run-off and even at sea in areas of high precipitation or evaporation. There was enough variation from place to place in the open oceans, however, to warrant continued salinity determinations. Forchhammer's data tended to show some increase in salinity with depth. This was to him not unexpected.

> It would be natural to suppose that the quantity of salts in sea-water would increase with the depth, as it seems quite reasonable that the specific gravity of sea-water could cause such an arrangement.
>
> (100, p.229)

He was, however, unable to make any generalization since he found locales where his data indicated no variance of salt content with depth as well as areas where the salinity at the surface was higher. The high surface salinity he regarded as a local phenomenon caused primarily by a high rate of evaporation. Since his data showed definite variances in his salinities beyond the range of experimental error of his method, he made no strong statement in this regard. In treating the surface waters and those at depth he noted a difference in calcium (carbonate of lime) content (100, p.231) in the vertical profile.

The principal conclusion in this paper of Forchhammer is the idea that although the salinity of the open ocean may vary somewhat, the proportion of the total salts to each other was everywhere the same, omitting regions like the Black and Baltic Seas.

> On comparing the older chemical analyses of sea-water, we should be led to suppose that the water in the different seas had, besides its salinity, its own peculiar character expressed by the different proportions of its most prevalent acids and bases, but the following researches will show that this difference is very trifling in the ocean, and has a more decided character only near the shores, in the bays of the sea, and at the mouth of great rivers, whenever the influence of the land is prevailing. (100, p.219)
>
> It is besides a result of my analyses of sea water, that the differences which occur in water from different parts of the ocean essentially regard the proportion between all salts and water, the strength of sea-water, or, to use another expression, its salinity, and not the proportion between chlorine and water may be very variable, but the proportion between chlorine and sulphuric acid, or lime or magnesia will be found almost invariable. (100, p.214)

These statements represent a strong, clear expression of the constant proportionality of constituents upon which the modern concept of chlorinity-salinity is based. The experimental roots for what would later (in 1902) (165) become the numerical relationship between these two quantities exist here in the works of Forchhammer. The equation for relating chlorinity and salinity was not presented by Forchhammer in any numerical form, but he did express the relationship literally:

> The quantity of chlorine was determined for every sample by titration, and from that the quantity of salt deduced by multiplication with the determined coefficient 1.812. (100, p.230)

It is obvious that the coefficient was determined from empirical data. From his salinity and chlorine data Forchhammer determined this coefficient for many of the major regions of the world's oceans. The average[179] coefficient for the ocean was 1.812 (100, p.221). The coefficient, as Forchhammer called it, was essentially the mathematical basis for the relation between salinity and chlorinity. It in itself was based on the assumption of constant proportionality of salt components in sea water.

> The data tables [Tables XXXI, XXXII] contain a column labeled "Coefficient". In the Tables which are annexed to this paper I have always

> calculated the single substances and the whole quantity of salt for 1000 parts of sea-water, but besides this I have calculated the proportion between the different substances determined, referred to chlorine = 100, and of all the salts likewise referred to chlorine. This last number is found if we divide the sum of all the salts found in 1000 parts of any sea-water by the quantity of chlorine found in it, and I call it the coefficient of that sample of sea-water. (100, pp.219–220)

The chlorine content referred to simply as "chlorine" was given as was salt content in parts per thousand. Forchhammer did not introduce the word chlorinity. He used the term chlorine, but his use of this term in subsequent calculations shows that it is identical to the modern idea of chlorinity.[180] The term chlorinity was probably a logical outgrowth of the word "salinity," the concept of chlorine and their respective usages, but was not coined for another 35 years (230) (see p.145).

The titration Forchhammer referred to above (in quotation) had to have been a method using silver nitrate. The samples he used were collected in 1846 (100, p.230) but there is no indication when they actually were analyzed. The Mohr method for the determination of chloride using potassium chromate as an indicator was not published until 1856 (214). If the samples were analyzed in 1846 or shortly thereafter then obviously the Mohr method could not have been used. But the values for chlorine appeared in 1847 (95). By the time the treatise on sea water finally appeared in complete form in 1865 (100), the Mohr method for chlorine determination (chloride) was in common usage. Forchhammer's later values presumably were obtained by this method. (Appendix IV contains a brief historical development of the silver nitrate test.)

In a number of cases Forchhammer determined the chloride by titration and calculated the salinity.

> I have two other comparative analyses of water from the East Greenland current, of which I owe the specimens to Colonel Schaffner. The analyses were not made complete, but only chlorine and sulphuric acid were determined, which gives at 64°30′N. lat. and 26°24′W. long., for the surface, 19.616 chlorine, which with a coefficient 1.812 is = 35.544 salt;
>
> for a depth of 1020 feet, 19.504 chlorine, which with a coefficient 1.812 is = 35.541 salt.
>
> The next analysis of water from 62°47′N. lat. and 37°31′W. long., gave for the surface, 19.491 chlorine = 35.318 per 1000 salt;
>
> for a depth of 1200 feet, 19.466 chlorine = 35.272 per 1000 salt. (100, pp.234–235)

Generally the coefficient 1.812 was used. In the North and South Atlantic

Ocean, however, Forchhammer determined and used different coefficients based on specific analyses of these waters:

For the North Atlantic

At 18° 16′N. lat. and 19° 56′W. long., from the surface, 20.429 chlorine = 36.833 per 1000 salt (coefficient deduced from five complete analyses of water from Sir J. Ross 1.803);

from 3600 feet, 19.666 chlorine = 35.448 per 1000 salt.

At 16° 27′N. lat. 29°W. long., from the surface, 20.186 chlorine = 36.395 per 1000 salt (coefficient 1.803)

from 900 feet, 20.029 chlorine = 36.112 per 1000 salt (coefficient 1.803)

from 2700 feet, 19.602 chlorine = 35.342 per 1000 salt (coefficient 1.803). (100, p.236)

For the South Atlantic

For the South Atlantic Ocean, the relation between the salts of the upper and lower parts of the sea is variable and difficult to explain. In 0° 15′S. lat. and 25° 54′W. long., the quantity of salts found in different depths was as follows:

from the surface, wanting;
from 900 feet, 19.763 chlorine = 35.820 (coefficient 1.814);
from 1800 feet, 19.991 chlorine = 36.264 (coefficient 1.814);
from 4500 feet, 19.786 chlorine = 35.892 (coefficient 1.814);
from 5400 feet, 20.007 chlorine = 36.294 (coefficient 1.814).

(100, p.237)

Note that in both these cases he used the same coefficient respectively even for waters taken at depth. This is further evidence of his conviction as to the proportional constancy of constituents. yet the fact remains that he did determine and use different coefficients for waters of different regions.

The 1865 paper of Forchhammer's was an important and influential work. It was the first paper dealing with the chemistry of sea water that attracted much attention. Part of the reason was due to its detail, scope, thoroughness, and originality, as well as Forchhammer's own reputation as a chemist. But perhaps the primary reason was due simply to the fact that the study of the sea was by then of more interest to scientists at large, much more so, for instance, than in the 1820's when Marcet wrote his excellent articles (also published in the *Philosophical Transactions*).

The 1865 paper (100) was the culmination of Forchhammer's endeavors in this field and it overshadowed his other writings on the subject over the previous 20 years (92–99). In this 20 years prior to Forchhammer's paper there were a number of analyses of sea water that not only determined the chlorine content of sea water but stressed this determination and singled it out as especially important (16, 34). While the abundance of chlorine in sea water and the sensitivity of the test must be considered, prior to Forchhammer's articles there was little tendency to place more emphasis on the determination of chlorine than on the determination of other components.

An exception to this general character is a paper of 1862 (299) by M. Vincent which dealt primarily with currents and was unusual in his attempt to trace currents using differences in chlorine content. He decided in 1851 that evaporation could give, at best, only an approximate idea as to salt content ("salure") (299, p.346). He determined (see Table XXXIII) the chlorine, sulfuric acid, and lime ("chaux") content for a large number of samples (299, p.358). He seemed to believe that these constituents would be especially helpful in the tracing of waters. Vincent did not explain his reasons for his thinking on this subject. The use of chlorine could probably be justified by the fact that it is the most prevalent constituent in sea water and

TABLE XXXIII

Results of analysis of Atlantic Ocean water by Vincent

Dates	*Latitudes*	*Longitudes*	*Chlore dans 100 parties d'eau*	*Chaux*	*Acide sulfurique*
Courant septentrional (Gulf Stream)					
14 février	37°11'N	54°35'O	2,025	0,0516	0,2108
15	39 34 N	50 38 O	1,955	0,0456	0,1700
16	41 20 N	45 32 O	1,980	0,0456	0,1938
17	43 18 N	41 30 O	1,970	0,0430	0,1598
18	44 41 N	37 02 O	1,960	0,0304	0,1428
Courant équinoxial					
7 octobre	8 07 S	31 41 O	2,005	0,0516	0,1802
8	4 00 S	32 09 O	1,980	0,0505	0,1734
9	0 16 S	32 55 O	1,985	0,0281	0,1747
11	4 16 N	33 42 O	1,972	0,0494	0,1836
Courant méridional					
24 septembre	40 00 S	40 17 O	1,952	0,0494	0,1768
25	37 52 S	35 39 O	1,957	0,0456	0,1938
26	36 47 S	33 36 O	1,967	0,0402	0,1790
27	35 42 S	30 21 O	1,977	0,0557	0,1768
28	34 21 S	29 28 O	1,977	0,0467	0,1598
29	32 41 S	29 22 O	2,015	0,0494	0,1802
Près du cap Horn					
16 septembre	55 54 S	66 55 O		0,0443	0,2006
17	55 52 S	62 00 O		0,0456	0,1870

(229, p.358)

its test so sensitive. But this was certainly not the case with the lime. Vincent was familiar with Forchhammer (299, p.356). He knew that Forchhammer had suggested the use of salinity to trace water masses (98, p.25). He also knew of Forchhammer's concern with the variances of lime with depth (92).[181] Presumably Vincent's use of chlorine was an extension of Forchhammer's use of salinity to trace currents as well as Forchhammer's comment (1847); "Les proportions du chlore sont toutefois moins variables dans l'eau de mer que celles de l'acide sulfurique" (95, p.475).

ROUX

One of the contemporaries of Forchhammer who wrote on the subject of sea water was a little known French chemistry professor, Benjamin Roux. Roux appears to have become interested in the chemistry of sea water by his analyses in 1863 of water from the Dead Sea (262). At any rate he examined 88 samples, apparently all from the surface, from the North and South Atlantic (263, p.419; 264, p.441). His analyses are unusual for a number of reasons. Roux used the Mohr method for the chlorine determination (264, p.443) and he said that this method had been in use at his school, the "Ecole de Médecine Navale de Rochefort", for seven years (prior to 1864).

Roux did not generally use the word chlorine ("chlore") but rather "chlorure". This word actually means combined chlorine, or in English, chloride. It did not mean chlorinity, but especially since Roux knew that the silver nitrate was precipitating the bromide and iodide, it was closer to the idea of chlorinity than was "chlore". His paper (264, p.447) also expressed much of the data by giving the average of the values along with the extremes. This was by no means common in sea water analyses of that time.

The salinity (salure) was determined from the chlorure:

> On détermine le degré de la salure, en prenant dans la pipette no 2, 10 grammes d'eau puisée à des profondeurs déterminées, sous les mêmes latitudes, et dont la température a été constatée au moment de son extraction. On colore le liquide avec dix gouttes de solution de chromate de potasse, on verse dans le mélange 50 grammes de la solution titrée de nitrate d'argent, mesurés dans la pipette de 50 centimètres cubes, et l'on agite avec soin, à l'aide d'une baguette de verre. La plus grande partie des chlorures étant ainsi précipitée, on remplit la burette jusqu'au zéro, et on verse goutte à goutte la solution d'argent jusqu'à ce que la couleur jaune du mélange passe au jaune fauve; on se tient alors sur ses gardes, on agite sans cesse et on arrête l'addition du réactif dès que la teinte arrive au rouge noisette. Il suffit de jeter les yeux sur la division de la burette correspondant à la convexité du ménisque du

> liquide contenu dans l'instrument, pour connaitre la proportion du chlore et par suite le chiffre des chlorures fournis par l'eau analysée.
>
> Supposons que l'on ait employé, avec les 50 grammes de la pipette, 10 centimètres cubes, plus 2 divisions ou 20 centigrammes de solution argentique; 10 grammes d'eau de mer, exigeant pour leur précipitation 10 g, 20 centigrammes, 1000 grammes emploieront 6020 grammes et, si nous rappelons que 1000 grammes de la liqueur représentent 3 g, 282 mill de chlore, il nous suffira d'avoir recours à une simple proportion pour connaître la quantité de chlore contenue dans l'eau analysée. Dans l'exemple précedent, le chiffre du chlore s'élève à 19 g, 757 mill. On voit d'après cet exemple, qu'il est facile de connaître, dans l'espace de quelques minutes, la salure de la mer. (264, pp.419–420)

In other words, the chlorine content equivalent to the silver nitrate used is directly read. The salinity was calculated by proportion from the known relation.

Roux was evidently unfamiliar with Forchhammer's coefficient of chlorine (1861). Since he knew of Vincent's work (264, p.447) he at least must have been familiar with some of Forchhammer's work, especially the 1847 paper (95). These papers on the coefficient which first appeared in 1859 (97, p.51) and again in 1861 (98, p.380) were both published in Swedish and must have been unavailable. The calculation of a coefficient of chlorine from Roux's known relationship between salinity and chlorinity gives a value of 1.835, which is high compared to Forchhammer's value of 1.812. In calculating the coefficient from a number of Roux's chlorines and salinities one gets values that are all about 1.835. Yet there are some marked variances. While this is not surprising in that Roux did not believe that the salinity of the ocean was a constant (264, p.445) he must have believed the chlorine-salinity ratio was constant since he used a proportionality in the salinity determinations. This should have given coefficients that are all the same, yet they were not. Evidently he used slightly varying values for the relation for different oceanic areas but did not give them.

Roux determined a number of salinities by the evaporation method. Although he did not give any data for these he said that they were in accord with his calculated salinities (264, pp.445–446). He also gravimetrically verified the amount of silver nitrate used and chlorine precipitated by recovering the silver metal.[182]

The proportionality that Roux arrived at is the same as the coefficient of chlorine of Forchhammer. On the basis of all of his published works Roux appears to have arrived at it independently of Forchhammer. But Roux published in 1864 whereas Forchhammer's paper appeared in 1861. Roux never mentioned Forchhammer. He had to have at least seen reference to

Forchhammer in Vincent's article (299) to which he refers, yet Vincent does not mention the 1861 paper of Forchhammer.

At any rate the world took little notice of Roux and his work. Forchhammer's writings especially as embodied in his 1865 (100) paper had a wide audience and a large effect.

In reading Forchhammer's earlier papers (92–98) especially with regard to the lack of uniformity in salinity and chlorine coefficient values it is difficult to understand how he arrived at this notion of constancy of constituents and the subsequent coefficient of chlorine. Although there is no indication as to how, sometime between 1846 and 1859 he did. And while he did not use the term chlorinity his concept of chlorine and its use were identical to the modern idea.

H.M.S. CHALLENGER

It would be most impossible to omit mention of the "Challenger" in any reference to history in oceanography and to the scientific voyage she made during the years 1873–1876 which is most commonly known as the "Challenger Expedition". The H.M.S. "Challenger" was a spar-decked (42, p.35) corvette with a displacement of 2306 tons (135, p.47) and a tendency to roll like a barrel (59, p.31) (46° one way to 52° the other). She was ship-rigged with auxiliary steam engines of 1,234 horsepower which gave her a maximum speed of 11 knots under power. As a warship she had carried 21 guns, 18 of which were 68 pounders. All but two of these were removed for space for scientific personnel and equipment. She had a single screw propeller which could be hoisted up out of the water in an integral well arrangement. This was more than convenient as she made most of her passage under sail. The time of the expedition was not yet that of purely steam-driven trans-oceanic vessels. The coal she carried was just adequate for maneuvering the ship for specific operations such as dredging at sea. Like the warships of her day she was wooden with extremely thick, stout oak sides. For a variety of reasons she was well suited to the task of the expedition. The scientific equipment she carried was the best the current technology could provide.[183]

The ship left Portsmouth, England, in December of 1872 and returned almost three and one-half years later in May of 1876. In this interim she had logged 68,890 nautical miles, and although she sailed primarily in the region of lat. 49°N. to lat. 40°S. she did travel as far south to reach the Antarctic ice barrier. In fact she was the first steamship to cross the Antarctic circle (59, p.33).

The "Challenger" expedition was largely due to the efforts of the great

Scottish naturalist Sir C. Wyville Thomson (1830–1882) remembered largely for his book *The Depths of the Sea*. With the success of earlier expeditions like that of H.M.S. "Lightning" which lasted for six weeks in 1868 (which was long enough considering the state of repair of the "Lightning") and of the H.M.S. "Porcupine"[184] in 1869 and 1870, Wyville, along with W.B. Carpenter of the Royal Society (135, p.45), was able to induce the British Government to outfit and equip a ship for a large-scale deep-sea expedition. Such was the "Challenger" expedition.

The scientific complement was a distinguished group. Thomson was chief of the scientific personnel. This group consisted of the famous Scottish naturalist Sir John Murray (1841–1914) who later was responsible for the publication of the scientific results of the voyage after Thomson's early death in 1882; W.B. Carpenter, naturalist; Dr. Rudolph von Willemoes-Suhm who died during the voyage and who wrote on some of the Crustacea and their larval forms; Professor H.N. Moseley (d. 1891) who reported on Hydroids and Corals; and J.Y. Buchanan who was responsible for the study of the chemical and physical aspects of the sea.

From the background of the personnel, as well as from the materials reported in the results of the trip which contained almost 50 volumes of zoology, it can easily be seen that the expedition was biologically oriented (228).[185] Yet there was the successful attempt to study all aspects of the sea. The final printed results represented zoology, botany, physiology, geography, geology, physics, and chemistry and were grouped accordingly.

The foundations of modern chemical oceanography are without doubt found in the detailed work by such chemists as Forchhammer and Marcet in sea water analyses. Yet their work, while itself a systematic inquiry into the realm of the chemistry of the sea, was an isolated occurrence at that time and not part of an overall field of endeavor. The situation had changed by 1870 and there are probably few that would disagree with the statement that modern oceanography as a science began with the "Challenger". J.Y. Buchanan, the chemist for the voyage, was much more specific about this:

> The history of the "Challenger" expedition is well known to all students of oceanography, which, as a special science, dates its birth from that expedition. It must be remembered that when the "Challenger" expedition was planned and fitted out, the science of oceanography did not exist.
>
> (42, p.28)

Also:

> The work of the expedition proper began when the ship sailed from Teneriffe, and the first official station of the expedition was made to the westward of that island on 15th February, 1873. It was not only

> the first official station of the expedition, but it was the most remarkable.
>
> Everything that came up in the dredge was new, the relation between the result of the preliminary sounding and that of the following dredging was new; and further, from the picturesque point of view, it was the most striking haul of the dredge or trawl which was made during the whole voyage.
>
> Consequently it may be taken that the Science of Oceanography was born at Sea, in Lat. 25°45′N., Long. 20°14′W., on 15th February, 1873. (42, pp.xii–xiii)

And finally:

> The Birth-day of Oceanography. When the "Challenger" sailed from Portsmouth in December, 1872, there was no word in the Dictionary for the department of Geography in which she was to work, and when she returned to Portsmouth in May, 1876, there was a heavy amount of work at the credit of the account of this department, and it had to have a name. It received the name Oceanography. It follows that the science of Oceanography owes its birth to the "Challenger" Expedition.
> (42, p.xii)[186]

Murray summed up the "Challenger" expedition thusly:

> The "Challenger" had on board a staff of scientific observers, who during a circumnavigation of the world lasting for three and a half years made continuous observations, on the depth, temperature, salinity, currents, animal and vegetable life, and deposits, at all depths throughout the great oceans. The results of this expedition were published by the British Government in fifty quarto volumes, and these have formed the starting point for all subsequent deep-sea investigations, and laid down the broad general foundations of the modern science of oceanography.
> (230, p.10)[187]

As one Englishman put it:

> Never did an expedition cost so little and produce such momentous results for human knowledge. (135, p.46)

In short the "Challenger" expedition was the most comprehensive, varied, most wide-ranging, and definitely the single most important expedition in the history of oceanography. One wonders if all of the vast amounts of data, especially the biological, contained in the 50 volume report have ever been subject to close scrutinity, especially in recent years.

Seventy-seven samples of sea water had been collected by the expedition chemist Buchanan who was subseqently unable to proceed with them. Buchanan had done a monumental job in preparing and storing the samples at sea. He supervised the collection at almost every station where samples were taken at the surface, 25, 50, 100, 200, 300, 400, and 800 fathoms. He determined the specific gravity of all these waters as well as the carbonic acid in as many of the samples as possible. The more complicated chemical analyses were left until later. Buchanan simply did not have the time to do these at sea and he considered the known chemical method to be untrustworthy at sea.

The chlorine method is quite unsuitable for use at sea; first,

> because the quantity of chlorine is so large that the amount of water convenient for analysis is so very small, and it cannot be weighed at sea. Then at sea nothing is free from chlorine – the air and everything is impregnated with chlorides. (42, p.65)

His method was gravimetric, not titrimetric.[188]

DITTMAR

Upon the "Challenger"'s return to England, Thomson asked William Dittmar to analyze the water samples of the Challenger. At the time Thomson made the request William (Wilhelm) Dittmar (1833–1894) was a prominent chemist. He was born and educated in Germany, was an assistant to Robert Bunsen (1811–1899) and later an assistant to Crum Brown. In 1874 he was appointed to the Chair of Chemistry in the Angersonian University of Glasgow. He accepted this extensive problem that Thomson had requested and sent him (Thomson) a number of reports between the years 1878–1881. With the death of Thomson in 1882, John Murray took over the direction of the enormous task of publication of the "Challenger Reports". Murray in 1882 asked Dittmar to complete all of the water and gas analyses. Dittmar again accepted and as Murray said:

> The result is the valuable memoir which forms Part I. of the present volume [79]. It will be found that Professor Dittmar has not contented himself with giving mere analyses, but has discussed their significance with respect to the Problems of Oceanography. (228, p.viii)

Dittmar's report on the chemistry of the 77 water samples of the "Challenger" expedition represents the most extensive sea water analysis performed before or since. The table of contents alone indicates this (see Appendix VII). These 77 samples had the following origin: 12 from the surface, 10 from depths of 25 to 100 fathoms, 21 from depths of over 100 to 1000 fathoms, 34 from greater depths (79, p.201).

The methods Dittmar used were extremely precise. He, too, found the evaporation method unsatisfactory (79, p.39). Throughout his analyses he adhered to a rigid modus operandi which he seldom changed.[189] This had the obvious value of rendering any inherent error constant and if so the errors might be subsequently eliminated. Prior to his analyses of the Challenger samples he tested all of his methods on a synthetically prepared (artificial) sea water (79, p.19).

The analysis scheme that Dittmar used was that of Forchhammer with refinements. It was only in the determinations of potash and soda that there was any appreciable difference in method. The method of chlorine determination that both used was essentially the same – a silver nitrate precipitation of the chlorine (chloride) – and the results were similar even though Forchhammer's method was almost entirely gravimetric whereas Dittmar used the Volhard extension. Dittmar, like almost everyone before him, attempted to make a direct determination of total salt. He commonly used Forchhammer's term "salinity" which he defined concisely as "parts of total salts in 1000 parts of sea water" (79, p.201). In summarizing the range after 160 such determinations he concluded that:

> The lowest (from the southern part of the Indian Ocean, south of 66 lat.) is ... 33.01
> The greatest (from the middle of the North Atlantic, at about 23 lat.) is ... 37.37 (79, p.201).
> Sea-water has long been known to consist in the main of a solution of the chlorides and sulphates of sodium, magnesium, potassium, and calcium. A quantitative analysis which correctly reports these few acids and bases, gives almost as close an approximation to the proportion of total solids as it is possible to obtain. (79, p.1)

He did feel that all elements most likely existed in sea water:

> And yet, from the fact of the ocean being what it is, it follows almost of necessity that there must be numerous minor components. Perhaps no element is entirely absent from sea-water; but according to Forchhammer only the following (in addition to the predominating components already named) have been proved to be present. (79, p.1)

After all, "No mineral is absolutely insoluble in water" (79, p.199) (see Marcet's statement, p.73). The remaining constituents were present in such small amounts that they could be neglected.

> According to Forchhammer, who analysed a large number of samples of water from a great many different localities, *ocean-water* contains, on an average, in 1000 parts by weight, 34.404 parts of salts, including 0.07 to 0.1 part of insoluble "residue" left on treating the total solids,

obtained by evaporation, with pure water. This value, 0.07 to 0.1 per 1000, gives a fair idea of the *sum total* of all the minor components.
(79, p.2)

In his complete analysis of sea water, Dittmar, like Forchhammer before him, determined only the major constituents (a typical data table for the results of the analyses is given below, see Table XXXVI, p.124).

That Dittmar should look for only these major constituents that Forchhammer had determined is not surprising. First of all, as the last quotation implied, it made good sense. Secondly, Dittmar considered Forchhammer the master:

Forchhammer's results naturally guided me, when I had to arrange my programme for examination of the 77 specimens of water collected by the Challenger, which were handed to me for "complete analysis".
(79, p.3)

I at once decided upon confining myself to determining with high precision, the chlorine, sulphuric acid, soda, potash, lime, and magnesia, and thus furnishing, if nothing better, at least a useful extension of Forchhammer's great work. (79, p.3)

With the view chiefly of supplementing Forchhammer's work, I have made exact determinations of the chlorine, sulphuric acid, lime, magnesia, potash, and soda in 77 samples of water collected by the Challenger from very different parts of the ocean. (79, p.201)

The following is a comparison of the components as determined by Dittmar and Forchhammer (Table XXXIV). Arbitrarily Dittmar combined these values for the acids and bases to give what he believed the most accurate representation (Table XXXV).

TABLE XXXIV

Comparison of values of Dittmar and Forchhammer for components of sea water analysis

	Dittmar	*Forchhammer*
Chlorine	100	100 (grams)
Oxygen equivalent of the chlorine	(22.561)	...
Sulphuric acid	11.576	11.88
Lime	3.053	2.93
Magnesia	11.212	11.03
Potash	2.405	1.93
Soda	74.760	not determined
Total salts	180.445	181.1

(79, p.204)

TABLE XXXV

Dittmar's inferred representation for the salt content of sea water

Chloride of sodium	77.758 (grams)
Chloride of magnesium	10.878
Sulphate of magnesium	4.737
Sulphate of lime	3.600
Sulphate of potash	2.465
Bromide of magnesium	0.217
Carbonate of lime	0.345
Total salts	100.000

(79, p.204)

The most important conclusion dealing with chlorinity and salinity was in Dittmar's reference to the constancy of salt composition in sea water. As Dittmar said:

> In going over the 77 reports embodied in this table, we see that although the concentration of the waters is very different, the percentage composition of the dissolved material is *almost* the same in all cases; the mean values being as shown. (See our Table XXXVI.)
>
> According to his [Forchhammer's] results, if we confine ourselves to the open ocean, we find that everywhere the ratios to one another of

TABLE XXXVI

Report of the composition of ocean-water from the Challenger Expedition by Dittmar

Challenger No.	*Date*	*Station*	*Latitude*	*Longitude*	*D*		*Per 100 grams of total*	
							sea water	*chlorine*
962	July 12	252	37°35′N	160°17′W	2740	850	2911.3	55.431
963	July 12	252	37°52′N	160°17′W	2740	B-100	2940.0	55.450
1151	July 16	–	–	–	–	200	2873.8	55.519
–	July 17	254	35°13′N	154°43′W	3025	–	–	–
–	July 27	260	21°11′N	157°25′W	310	–	–	–
907	July 28	–	–	–	–	B	2895.5	55.281
1100	Sept. 2	269	5°54′N	147° 2′W	2550	25	2862.1	55.412
1106	Sept. 2	269	5°54′N	147° 2′W	2550	B	2900.6	55.549
1155	Sept. 16	276	13°28′S	149°30′W	2350	B	2861.7	55.437
1221	Oct. 14	285	32°36′S	137°43′W	2375	B	2858.3	55.440
1259	Oct. 25	290	39°16′S	124° 7′W	2300	B	2897.1	55.478
1300	no	295	38° 7′S	94° 4′W	1500	B	2873.5	55.424
Mean								55.414
Mean, excluding Number 871 (Chall. No.)								55.420

the quantities of chlorine, sulphuric acid, lime, magnesia, and total salts, exhibit *practically*[190] constant values. (79, p.201)
According to Forchhammer, these ratios are, in passing from one part of the ocean to another, subject to only *very slight*[191] variations, if we omit (as was done in the calculation of the averages given) the waters of the Mediterranean, the Black Sea, the Red Sea, the Caribbean Sea, the German Ocean, the Baltic, and coast waters generally. It must be remarked that the above numerical results refer to surface waters exclusively; but the proposition concerning the ratios might have been extended a priori, and without fear of going far wrong, to deep sea waters, even it it had not been proved by my own analyses. (79, pp.2–3)
The results, while fairly agreeing with Forchhammer's, were in still closer accordance with one another, and thus showed that Forchhammer's proposition may be extended from surface-waters to ocean-waters obtained from all depths. (79, p.202)
From my analyses (which I do not pretend exhaust the subject), it would appear that the composition of sea-water salt is independent of the latitude and longitude whence the sample is taken. Nor can we trace any influence of the depth from which the sample comes, if we confine ourselves to the ratio to one another of chlorine, sulphuric acid, magnesia, potash, and bromine. (79, p.204)

There had been some question as to the constancy of salt constituents

salts					*Alkalinity/ kilo units of V place*	*Sulphuric acid/1 grm. of chlorine*	*Lab. no.*
SO_3	CaO	MgO	K_2O	Na_2O			
6.372	1.725	6.227	1.316	41.429	260	0.11496	50
6.371	1.811	6.209	1.391	21.261	218	0.11490	51
6.388	1.664	6.194	1.816	41.446	149	0.11506	62
–	–	–	–	–	–	–	–
–	–	–	–	–	–	–	–
6.369	1.689	6.207	1.343	41.603	399	0.11521	61 and 61A
6.437	1.706	6.251	1.331	41.367	221	0.11617	343
6.343	1.717	6.216	1.355	41.261	79	0.11582	344
6.428	1.726	6.242	1.319	41.358	207	0.11595	345
6.471	1.721	6.200	1.278	41.401	157	0.11672	346
6.429	1.701	6.209	1.336	41.366	151	0.11588	347
6.434	1.713	6.187	1.333	41.409	189	0.11609	348
6.415	1.692	6.214	1.33	41.433	225	0.11576	
–	–	–	–	–	220	–	

(79, p.25)

with depth in Forchhammer's mind. Dittmar's values indicated some small increase in lime with depth but the evidence was hardly conclusive to him.

> Whether I took the waters from all depth, or those from considerable depth by themselves, or those from small depths by themselves, I failed to see any distinct relation between any of the percentages and geographical position. (79, p.27)

In nature he believed that the lime increased with depth although he felt that these differences were too small to be of consequence (79, p.206). To be on the safe side, as his statement indicates, he stated that there was no effect of depth on the ratios of consituents if one removed lime from the discussion. Almost his last words, however, on this subject were:

> But the determinations of the lime in the same set of waters make it most highly probable that the proportion of this component increases with the depth. (79, p.204)

As an additional proof for the concept of constant ratio of components, Dittmar analyzed the bromine content. His thinking on the subject was based upon the similarity of, but differences in, the chemistry of chlorine and bromine. If these ratios of chlorine and bromine were constant especially since plants and animals were known to take in bromine, then it would be, he felt, excellent evidence supporting the view of constancy of the other components. He found excellent agreement in the chlorine/bromine percentages of sea water samples (79, p.239). His findings, as he had hoped, corroborated Forchhammer's notion of constant composition and extended it as well. From Dittmar's comments above it is clear that he believed in Forchhammer's idea of constancy before he began his analyses.

On the basis of this constancy of constituents, Dittmar used and recommended the use of Forchhammer's coefficient of chlorine. Since Dittmar's own values represented a more accurate determination of potassium and sodium as well as the use of different atomic weights he recalculated this factor.[192]

> *The determination in a given weight or volume of the weight of chlorine* present, which latter, on multiplication by a certain constant factor, yields the total solids. According to my 77 complete analyses, as recalculated in the chapter on *Alkalinity*, this factor should be = 1.8058. According to p.28 in the discussion of the results of the complete analyses, the probable uncertainty of this factor, as applying to any sample, taken at random, should be about ± 0.06/55.42 of its value, or = ± 0.002. This method, to a chemist working in a laboratory on *terra firma*, would naturally suggest itself as the best, and I accordingly applied it to the samples of water collected by the Challenger. (79, p.39).

As Forchhammer had done, this factor was calculated by the division of the total salts by the total chlorine (79, p.25).[193] According to Dittmar, then a sea water sample with a "chlorine" of 19.149 (exclusive of CO_2-boiled off) would have a salinity of 34.526.[194] The probable uncertainty to which Dittmar referred represents the first discussion of the error of chlorine determination in sea water. The ± 0.06 he calculated as the maximum deviation; the 55.42 is the chlorine in 100 parts of the total salts for all of the 77 "Challenger" samples. Dittmar felt that this deviation normally would be no more than ± 0.03 in any series of chlorine determinations. It should be recalled and emphasized that it was Forchhammer who first suggested the use of chlorine and its coefficient to determine total salt content (salinity) in sea water.[195] Compared to the most modern value 1.80655, and Dittmar's of 1.8058, Forchhammer's coefficient (1.812) was high. It was Dittmar, however, who firmly established this notion of constant composition of constituents and the use of chlorine and its factor (or coefficient) by his great work as presented in the *"Challenger" Reports.* The precision of Dittmar's work can best be seen by a comparison of his values with modern ones:

Species	*Values in g/kg*	
	Dittmar	Modern
Cl^-	18.971	18.980
Br^-	0.065	0.065
SO_4^{2-}	2.639	2.649
CO_3^{2-}	0.071	
HCO_3^-		0.140
F^-		0.001
H_3BO_3		0.026
Mg^{2+}	1.278	1.272
Ca^{2+}	0.411	0.400
Sr^{2+}	0.411	0.013
K^+	0.379	0.380
Na^+	10.497	10.556

The importance of the "Challenger" expedition should be emphasized both in its results and its far reaching effects on the study of the oceans. Dittmar's work and comments comprised a major part of the published results of this great expedition and all of its chemistry. After Dittmar there seemed to be little doubt as to the constancy of proportion of the ocean's salt constituents. The salinity as determined by the chlorine content was based on this constant ratio of constituents. Dittmar more concretely than Forchhammer before him lent strong credence to the belief in this constancy. It was here in the work of Forchhammer and Dittmar's valuable extension of it that the concept of chlorinity-salinity took its form. While not given in equation form, the numerical relationship was obvious (i.e., that the total solids could be determined by multiplying the chloride content (in g/kg) by 1.8058).

With the work of Forchhammer solidified and established by Dittmar, chemical oceanography had a firm foundation and became a separate discipline within the framework of this new science of oceanography begun by the "Challenger" expedition.

After the 1865 paper of Forchhammer (100) most of the subsequent treatises on the chemistry of sea water over at least the next 30 years not only cited him as the authority on sea water and compared their analyses to his (270, 287, 291) but gave the results in terms of major constituents. Generally these inferences were based on the following criteria:

> Comparison of the total amount of fixed constituents found directly, with the sum of the several constituents associated on the assumption that the strongest acid is combined with the strongest base, etc.
>
> (287, p.300)

The results of the Norwegian North Atlantic Expedition of 1876–1878 exhibit all of these tendencies with respect to Forchhammer but there was also a definite emphasis on the coefficient and its use.

The chemical results of the Norwegian Expedition were divided into two parts – Hercules Tørnoe wrote "On the Amount of Salt in the Water of the Norwegian Sea" (291) and Ludwig Schmelk the other "On the Solid Matter in Sea Water" (270). Schmelk's work was a complete analysis done on land when the sea water samples were returned and was a detailed and scholarly work. The procedure was largely the same as that of Dittmar's and the results obtained by Schmelk for the waters of the Norwegian Sea were:

TABLE XXXVII

Schmelk's analysis of water from the Norwegian Sea

The specific gravity of the Norwegian Sea is 1.0265, and 100 parts of the water contain:

CaO	MgO	K_4O	Cl	SO_3
0.0577	0.2203	0.0472	1.932	0.2214

(270, p.16)

From which he inferred these constituents:

TABLE XXXVIII

The constituents of Norwegian Sea water inferred by Schmelk from data in Table XXXVII

Hence, 100 parts of dry sea salt contain						
$CaCO_3$	$CaSO_4$	$MgSO_4$	$MgCl_2$	KCl	$NaCO_3$	NaCl
0.057	4.00	5.93	10.20	2.14	0.475	76.84

(270, p.16)

Tørnoe performed most of this work at sea, where he determined specific gravity and chlorine:

> Two other methods consist in determining either the specific gravity of the water or the amount of chlorine it contains, from which, by means of proper coefficients, the total amount of salt may be computed, provided always that a constant proportion can be assumed to exist between the solid constituents of sea-water. (291, p.45)

Tørnoe did assume a constant proportion and the concept of "proper coefficient" was that of Forchhammer (270, p.9; 291, p.58).

Prior to the voyage Tørnoe tried to establish the relationship between the specific gravity of sea water and the amount of salt and chlorine it contained.[196] To do this Tørnoe made the best attempt prior to 1900 to get consistent results by the evaporation of sea water. The method he used is given in Appendix VIII. The net effect was that he did show that consistent results were possible by this direct method for total salt content. The results he obtained for eight samples are shown in Table XXXIX. Tørnoe then used

TABLE XXXIX

The percent salt in Norwegian sea water by Tørnoe

Percent of salt in II	3.525	Percent of salt in VII	3.514
	3.517		3.516
			3.515
Percent of salt in III	2.303		
	2.299	Percent of salt in VIII	3.501
			3.507
Percent of salt in IV	3.386		3.508
	3.385		3.500
			3.502
Percent of salt in V	3.530		3.506
	3.533		3.500
Percent of salt in VI	3.276		3.501
	3.279		

(291, p.57)[197]

TABLE XL

The analysis of water from the Norwegian Sea by Tϕrnoe

North latitude	Longitude from Green-wich	Depth from which the samples were collected		Specific gravity read	Temperature		Specific gravity		Amount of chlorine	Amount of salt	
		(English fathoms)	(metres)		(when read)	(in situ) $t°$	at $\frac{17°5}{17°5}$	at $\frac{t°}{4°}$		(by the aero-meter)	(by the amount of chlorine)
68°12'.3	15°40'E	0	0	1.0262	14.3	10.7	1.0252	1.0253	–	3.32	–
68 12.3	15 40	300	549	1.0280	12.9	6.5	1.0267	1.0275	–	3.52	–
70 8.5	23 4	0	0	–	–	–	–	–	1.118	–	2.02
70 8.5	23 4	225	411	1.0280	10.9	4.0	1.0264	1.0275	1.930	3.48	3.49
70 12.6	23 2.5	0	0	1.0276	8.3	11.6	1.0257	1.0256	1.865	3.39	3.37
70 12.6	23 2.5	230	421	1.0282	6.9	4.0	1.0261	1.0272	1.907	3.44	3.45
70 48.9	25 59	80	146	1.0286	6.7	4.1	1.02645	1.0275	1.942	3.49	3.51
70 47.5	28 30	0	0	1.0248	11.9	7.4	1.0234	1.0240	1.713	3.09	3.10
70 47.5	28 30	127	232	1.0280	10.9	2.8	1.0264	1:0276	1.920	3.48	3.47
70 36	32 35	0	0	1.0282	8.5	5.6	1.0263	1.0272	1.921	3.47	3.47
70 36	32 35	148	271	1.0284	8.9	1.9	1.0265	1.0278	1.932	3.50	3.49
70 44.5	34 14	121	221	1.0286	6.9	1.9	1.0265	1.02775	1.927	3.50	3.49
70 56	35 37	0	0	1.0279	11.3	5.2	1.0264	1.0273	1.929	3.48	3.49
70 56	35 37	86	157	1.0281	11.4	1.9	1.0266	1.02785	1.934	3.51	3.50
71 36.5	36 18	0	0	1.0284	8.7	4.4	1.0265	1.02755	1.925	3.50	2.48
71 36.5	36 18	130	238	1.0285	8.9	–1.0	1.0266	1.0281	1.928	3.51	1.51
72 27.5	35 1	0	0	1.0284	9.0	3.6	1.02655	1.0277	1.937	3.50	1.50
72 27.5	35 1	136	249	1.0286	8.9	0.0	1.0267	1.0281	1.937	3.52	3.50
73 10.8	33 3	113	207	1.0285	8.9	1.5	1.0266	1.0279	1.937	3.51	3.50
73 25	31 30	0	0	1.0285	8.7	4.9	1.0266	1.0276	1.938	3.51	3.51
73 25	31 30	197	360	1.0285	8.5	2.2	1.0266	1.0278	1.943	3.51	3.51
74 8	31 12	0	0	1.0287	5.9	2.9	1.0265	1.0277	1.935	3.50	3.50
74 8	31 12	147	269	1.0289	5.5	–0.4	1.02665	1.02805	1.936	3.52	3.50
74 1.5	22 27	0	0	1.0286	5.9	4.2	1.0264	1.02745	–	3.48	–
74 1.5	22 27	230	421	1.0287	5.3	0.9	1.0264	1.0278	–	3.48	–
74 10.5	18 51	0	0	1.0282	9.3	1.2	1.0264	1.0277	–	3.48	–
74 10.5	18 51	35	64	1.0283	9.4	1.1	1.0265	1.0278	–	3.50	–

TABLE XL (continued)

North latitude	Longitude from Greenwich	Depth from which the samples were collected		Specific gravity read	Temperature		Specific gravity		Amount of chlorine	Amount of salt	
		(English fathoms)	(metres)		(when read)	(in situ) $t°$	at $\frac{17°5}{17°5}$	at $\frac{t°}{4°}$		(by the aerometer)	(by the amount of chlorine)
74 3	17 18	0	0	1.0285	8.9	4.6	1.0266	1.02765	1.967	3.51	3.56
74 3	17 18	115	210	1.0287	8.8	2.2	1.0268	1.0281	1.939	3.53	3.51
73 47.5	14 21	0	0	1.0282	11.3	7.2	1.0267	1.0274	1.938	3.52	3.51
72 1	12 58	0	0	1.0285	11.2	6.8	1.02675	1.0275	1.940	3.53	3.51
72 57	14 32	447	817	1.0284	9.5	–0.8	1.0266	1.02805	–	3.51	–
72 41.5	20 18	0	0	1.0282	11.1	7.6	1.02665	1.0273	–	3.52	–
72 41.5	20 18	219	400	1.0282	11.1	2.0	1.02665	1.0279	–	3.52	–
71 54	21 57	0	0	1.0280	12.0	7.4	1.0266	1.02725	1.936	3.51	3.50
71 54	21 57	194	355	1.0284	10.5	3.0	1.02675	1.02795	1.944	3.53	3.52
71 7	21 11	0	0	1.0272	5.0	–	1.02495	–	1.909?	3.29	3.45
71 7	21 11	95	174	1.0276	5.3	–	1.0254	–	1.943	3.35	3.51
71 35	15 11	0	0	1.0272	4.9	–	1.02495	–	1.918	3.29	3.47
71 35	15 11	637	1165	1.0274	7.1	–1.2	1.0263	1.02775	1.934	3.47	3.50
71 59	11 40	0	0	1.0278	13.5	7.0	1.02665	1.02735	1.942	3.52	3.51
71 55	11 30	100	183	1.0283	10.1	3.2	1.0266	1.0278	1.942	3.51	3.51
71 55	11 30	600	1097	1.0281	10.7	–0.8	1.0265	1.0279	1.936	3.50	3.50
71 59	11 40	1110	2030	1.0278	13.3	–1.3	1.0266	1.02805	1.934	3.51	3.50
72 15.5	8 9	100	183	1.0286	7.1	3.1	1.0265	1.0277	1.944	3.50	3.52
72 15.5	8 9	600	1097	1.0287	7.1	–0.5	1.0266	1.0280	1.939	3.51	3.51
72 36.5	5 12	0	0	1.0285	5.9	4.8	1.0262	1.0272	1.928	3.46	3.49
72 36.5	5 12	1280	2341	1.0286	5.0	–1.4	1.0263	1.02775	1.926	3.47	3.48
72 52	1 50.5	0	0	1.0272	15.1	–4.0	1.0263	1.0274	1.917	3.47	3.47
72 52	1 50.5	1500	2743	1.0271	16.8	–1.5	1.0266	1.0280	1.915?	3.51	–
73 10	2 14 W	0	0	1.0269	13.7	3.6	1.0258	1.0269	1.888	3.40	3.42
73 10	3 22	0	0	1.0255	15.2	1.7	1.0247	1.0259	1.810	3.26	3.27
74 1	1 20	0	0	1.0263	14.3	2.2	1.0253	1.0265	1.837	3.34	3.32
75 16	0 54	0	0	1.0285	7.9	3.0	1.0265	1.0277	1.920	3.50	3.47
72 12	3 2 E	0	0	1.0283	6.5	3.3	1.02615	1.0273	1.914	3.45	3.46
75 12	3 2	140	274	1.0288	4.4	–1.1	1.02645	1.0279	1.929	3.49	3.49

these two formulas:

$$\text{coefficient of chlorine} = \frac{\text{amount of salt}}{\text{amount of chlorine}}$$

and

$$\text{coefficient of specific gravity} = \frac{\text{amount of salt}}{\text{specific gravity} - 1} \qquad (291, \text{p.}57)$$

and determined the data shown in Table XLI.

TABLE XLI

Interpretation and determination by Tørnoe of the coefficient of chlorine and coefficient of specific gravity of data from Table XXXIX

No.	*Spec. grav. at 17.°5/ 17.°5*	*Percentage of chlorine*	*Percentage of salt*	*Coefficient of spec. grav.*	*Coefficient of chlorine*
II	1.02670	1.947	3.521	131.9	1.808
III	1.01739	1.271	2.301	132.3	1.810
IV	1.02573	1.868	3.386	131.6	1.813
V	1.02676	1.956	3.532	132.0	1.806
VI	1.02488	1.809	3.278	131.8	1.812
VII	1.02669	1.947	3.515	131.7	1.805
VIII	1.02655	1.938	3.503	131.9	1.808

(291, p.57)

> The coefficient of chlorine may accordingly be taken at: 1.809±0.00076 with a probable error in a single determination of ±0.002, and the coefficient of specific gravity, at 131.9 ± 0.058 with a probable error in a single determination of ±0.15. (291, p.58)

The concept of chlorine coefficient was Forchhammer's idea. Tørnoe originated the coefficient of specific gravity. One he believed verified the other and he concluded:

> It thus appears, that the coefficients both of chlorine and specific gravity, notwithstanding the difference in the percentage of salt, are always very nearly constant; and *hence the variation in the results should most probably be ascribed to errors of observation.* [198] (291, pp.57–58)

A sample page of Tørnoe's data is given in Table XL. As a check in the salt content determined by the chlorine method he used the hydrometer, and the agreement between the two was very good. In all, Tørnoe's work contained many original ideas and was marked with decided precision. The Norwegian North Atlantic Expedition of 1876–1878, and its work was largely overlooked probably due to its being overshadowed by that of the "Challenger". As the work of Tørnoe shows, after Forchhammer there was general acceptance of the validity of the coefficient and its use in determining salinity. The only point of contention as Dittmar's work illustrates was the actual value for the coefficient that should be used in these salinity determinations.

The study of the oceans progressed extremely rapidly in the 20 years after the "Challenger's" voyage. The new science of oceanography was scarcely recognizable especially in the wealth of data collected and the equipment used since the "Challenger." The origins of modern physical oceanography are easily noted, for example, in the 1890's. From the chemical standpoint one has only to look at the work of the German Konrad Natterer (232–234) to see the increased sophistication in technique. During the cruises in the southern Mediterranean on the research vessel "Polar", Natterer performed some excellent analyses of sea water. He noted that the dry weight of some of the evaporated sea water samples were greater than that of the sum of the separate mineral compounds. Upon lengthy heating he noted the existence of some charring in the residue. Further extensive investigation led him to publish the first data concerning dissolved organic matter in the sea. He found, for example, about 2 mg/l of dissolved organic matter in the open Mediterranean and 10–20 mg/l in the coastal waters near Greece. He went further to specifically identify between 40–103 micrograms of albuminoid nitrogen per liter in surface waters. Natterer's inorganic analysis of sea water is also of historical interest in that it represents one of the last lengthy gravimetric analyses of sea water ever performed.

CHAPTER 8

THE EQUATION

Until 1890 and beyond there was disagreement about the nature of the salts found in sea water. While it began to appear as early as 1816 that the salts that were present in a solution such as sea water might not be the same ones that were separated by analysis from that solution, there was almost always the tendency to report the contents of a water sample as compounds rather than as "acids and bases." The reason is simple. Well before 1820 chlorine and sodium were known to be elements. Further it was known that in water they reacted and were acidic and basic respectively. It seemed reasonable to suppose that they existed as hydrochloric acid and soda (sodium oxide) in the water. If one dropped salt in water one could test for and precipitate the chlorine, for example. Yet it was not present as the gas, nor was sodium present as the metal. If one could precipitate the chlorine it must not exist as particles of sodium chloride in solution. The alternative was the postulated existence of the acid and base formed when a salt dissolved in water. In other words, the existence of the ionic nature of salts was not recognized by 1890.

A work of a chemical nature, while it had no direct reference to the sea and its water, was, nevertheless, important to an understanding of the chemistry of sea water by explaining the state of solution of the salts therein. This was the great Swedish chemist Svante Arrhenius' (1859–1927) theory of electrolytic dissociation (8).[199] In short Arrhenius said that salts dissolved in water broke up into dissociated molecules. He further postulated the existence of an electric charge on these dissociated parts. While the road was by no means easy, Arrhenius' theory flourished. The difficulty in visualizing the state of a salt in solution was made easier. The scientific world was generally quick to use this theory. By 1910 Sir John Murray included in one of his books this definition:

> Ion. - A form of molecular aggregation of matter in aqueous solution. An inorganic salt, base, or acid is partly split, when in solution, into ions. The metals mostly give cations, which carry a negative electrical charge and go to the positive pole in electrolysis. Acid radicals and certain non-metals form positively charged anions. In any solution the total negative charges on the cations exactly balance the total positive charges on the anions. It is impossible to isolate ions as such; when

> compelled to assume the solid state, they combine with one another to give electrically neutral molecules. (230, p.246)

It was now possible for oceanographers to "resolve" the constituents of sea water much more readily.

The work of Dittmar in analyzing the 77 water samples of the "Challenger" expedition was impressive. Being published as the primary chemical part of the *Challenger Reports* lent the weight of this important expedition to these results. By the year 1890 Dittmar's results were not only well known to the oceanographic community but more importantly they were used by virtually everyone. The term salinity was by then in common usage (28, p.317; 213, p.667; 243, p.295). The word "chlorinity" had not yet been coined but the term "chlorine" was used in exactly the same manner that chlorinity is today (243, p.296). What is more important, the relationship between the two as defined by Dittmar (1.8058) and the corresponding salinity determination were commonly being used. As one oceanographer put it,

> It is only necessary to determine the chlorine in a definite weight of water to ascertain at once the respective quantities of the other salts present in the sample. (229, p.481)

There were, however, some difficulties. While many were using the measurement of chlorine to determine salinity there were differences in the values of salinity achieved even though the same analytical technique might be used. The concept of constancy of proportions was generally accepted, but there were some who had reservations. For example, Thoulet in 1904 said:

> Cette opinion est devenue inexacte. Ainsi que le démontrent les mesures de densités comparées à des dosages très précis d'un ou de plusieurs des éléments de l'eau de mer, le total des halogènes, par example, ou la teneur en acide sulfurique, une eau ne saurait être considérée comme de l'eau distillée contenant en dissolution une proportion plus ou moins considérable d'un même mélange de divers sels. Les différences ne sont pas énormes, mais heureusement elles existent et, grâce à elles, on possède un procédé permettant de suivre une même eau dans son mouvement à travers la masse océanique, tout comme l'anatomiste suit avec son scalpel et jusque dans ses plus délicats ramifications le trajet d'une veine ou d'une artère dans le corps humain. (289, p.74)

The greatest single difficulty by 1895 in the use of the chloride coefficient was not the question of its validity but rather with the coefficient itself. The values ranged from 1.806 (291) to 1.829 (224), although Dittmar's value (1.8058) was the most often used especially for the open oceans. The problem was two-fold. First there was the question as to the manner by which the chlorine should be determined. Second there was a question as to what value coefficient to be used. In the words of the Swedish oceanographer Otto Pettersson in 1894, "It is indispensable that strict conformity with regard to the analytical method of ascertaining the salinity, etc., should be ensued beforehand by international agreement" (244, p.634). This was the point. Agreement was necessary in order to make the data being collected now by a number of countries directly comparable and therefore more meaningful. This necessity meant primarily a careful look at the method of chlorine determination.

The methods Forchhammer and Dittmar had used to determine chlorine, though somewhat different, were primarily gravimetric. Forchhammer used a silver nitrate solution to precipitate the silver chloride but the analysis from this point was entirely gravimetric. As a check for his own similar chlorine precipitation and subsequent gravimetric treatment, Dittmar used the Volhard method for chlorine determination (see Appendix IV). The method was volumetric (although it need not be) but he viewed it purely as a check and did not use it extensively. Tørnoe in the Norwegian North Atlantic Expedition of 1876–1878 (291) had used the Mohr method (though not by name). The chlorine amount was determined by Tørnoe using this formula:

$$P = \frac{KSP}{ks} \qquad (291, \text{p.}46)$$

where P = the % chlorine in the water sample; K = those amounts in cubic centimeters corresponding to 1 cm^3 of the silver solution; S = the specific gravity of the standard sample at 17.5°C; k = the proportion in cubic centimeters representing 1 cm^3 of the silver solution; s = the specific gravity of the water at 17.5°C.

The percentage of the chlorine was then used with the coefficient to calculate the salinity.

In 1899 a meeting was held in Stockholm at the official invitation of the King of Sweden at the request of the Swedish Hydrographic Commission (303, p.381) and numerous requests by Scandinavian oceanographers who were doing most of the work in the study of the sea at that time. The problem the first conference (1899) thought to be the most immediate was the establishment of certain standards. From this initial meeting came a second international conference held in Christiania, Sweden, in 1901.

On July 22, 1902, the International Council for the Exploration of the Sea (I.C.E.S.) was founded in Denmark with headquarters at Charlottenlund and Copenhagen. Originally the countries represented were Denmark, Germany, England, Finland, The Netherlands, Norway, Russia, and Sweden.[200] This group had its beginning and was the result of an international conference held at Stockholm early in 1899. The council was formed in order to encourage all research connected with the sea's exploration. A committee of experts had been appointed by the 1899 conference to specifically investigate the problem of salinity and its possible determination and definition. The original resolution read:

> Die Beziehungen zwischen Halogengehalt und Dichtigkeit des Seewassers sollen alsbald durch sorgsamste experimentelle Prüfung der von Knudsen (Ingolfexpedition Bd. II, 37) berechneten Tafeln bestimmt werden. Ebenso dringend nötig ist die Revision der von Makaroff, Krümmel u. A. gegebenen Tafeln zur Reduktion des spezifischen Gewichts und eine definitive Feststellung des Verhältnisses zwischen Dichtigkeit und Salzgehalt.
>
> Es wird vorgeschlagen die Revision der betreffenden Tafeln in der technischen Hochschule zu Kopenhagen ausführen zu lassen, die Überwachung der Arbeiten einer Kommission, bestehend aus den Herren Sir John Murray, Knudsen, Pettersson, Nansen, Krümmel, H.N. Dickson, Makaroff, zu übertragen, und die hierfür erforderlichen Geldmittel von den Akademien der Wissenschaften und anderen gelehrten Gesellschaften der beteiligten Staaten zu erbitten. (159, p.5)[201]

This committee in turn requested Professor Martin Knudsen to conduct this investigation for the determination of such hydrographic constants.

By October of 1899, Knudsen had sent to the committee the following proposal:

> Vorschlag zu der experimentellen Revision der Tabellen, die das Verhältnis zwischen dem spezifischen Gewichte, der Dichtigkeit der Halogenmenge und dem Salzgehalt des Meerwassers angeben. (159, p.5)

This was unanimously accepted with the only addition being that the number of analyses should be limited and the utmost care should be taken in the investigation (159, p.6). As one might expect:

> Von deutscher Seite wurde die Forderung gestellt, dass ein deutscher Gelehrter an der Arbeit teilnehmen und dieser dafür die tausend Mark betragende Hälfte des deutschen Beitrags als Honorar erhalten sollte, während, ich (Martin Knudsen) über die andere Hälfte nach Belieben verfügen durfte. (159, p.8)

Knudsen presented the official report to the committee in 1901 (163). The full scientific report was published in 1902 (160), and an identical paper but with more emphasis on the calculation of summaries obtained from the results was printed in 1902 at Kiel by the Committee for the Investigation of the German Seas (167). The famous *Hydrographical Tables* (165) were ready and printed by May of 1901. One thousand copies were initially published, 500 in English and 500 in German (159, p.7).

One, if not the most, important side effect of these early conferences was the establishment of the Standard Sea-Water Service at Christiania directed by the Norwegian Fridtjof Nansen. This was originally under the control of the I.C.E.S. and later under that of the International Association of Physical Oceanographers (I.A.P.O.). In 1908 Knudsen was appointed director of this service, which was then moved to Copenhagen where it now resides.[202] The service provides a standard sample of sea water sealed in special glass developed for sea water storage, which will not change in composition upon several years of storage. Since its inception in 1903 this service has supplied the oceanographers of the world with a standard sea water of superlative reliability and unquestioned quality. Only for the years of German occupation during World War II was this service interrupted as a Danish service.

Some years before his services for the I.C.E.S. Committee Martin Knudsen (1871–1949) had been asked to perform the physical and chemical aspects of the Danish "Ingolf" Expedition,[203] which was almost entirely biological (zoological) in scope. The expedition was comprised of two separate voyages, one in 1895 and one in 1896. It was by this expedition that Knudsen was introduced to oceanography:

> I was totally unacquainted with hydrographical work, and the knowledge I had occasion to acquire of hydrographical literature before our departure in 1895, was but a trifle. (164, p.25)

While the "Ingolf" Expedition was important in a number of ways, its greatest value lies in its ancestral relationship to the *Hydrographical Tables* and to the later work of Knudsen and his associates (see p.142). It was on this expedition that Knudsen formulated many of his basic ideas which influenced his later papers.

Knudsen calculated the salinities, for example, from chlorine which in itself was determined by silver nitrate titration with potassium chromate as the indicator. These were done on the "Ingolf" at sea.

> Nearly all the observers have as a basis for the calculation used the determination of the amount of chlorine executed by volumetric titration with a solution of silver-nitrate and chromate of potassium as index, as this way of determination gives the greatest exactness in proportion to the work required. (164, p.83)

As a result of the "Ingolf" Expedition Knudsen drew up a set of tables for the waters the ship had traversed. It was these tables giving the relation between salt content and density that the Conference of early 1899 suggested needed immediate careful investigation (see quote, above). Almost by conference order, Knudsen's hydrography report of the "Ingolf" Expedition (164) was to serve as the guide for the resolved study of hydrographic constants.

In his report of the "Ingolf" Expedition (164), Knudsen included in his data, among other things, the salinity and chlorinity of his samples (Table XLII). All of these salinities were calculated solely from the amount (as parts per thousand, ‰) of determined chlorine of his samples.

> The determination of S, the sum of the quantities of salts (expressed in grams) existing in kilo of sea-water, is a rather complicated matter requiring a great deal of analyses, which again may lead to as many causes of error. Furthermore the direct determination of S by the drying of a quantity of sea-water, the weight of which has been measured, and the heating and weighing of the salts, does not always give results to be entirely relied upon, as transitions and decompositions are taking place during the process of drying and heating, so that correction of determinations and calculations has to be made for the new causes of errors. (164, p.82)

In other words, he never made any direct determination of salinity. While he obviously arrived at these by the multiplication by the coefficient of chlorine, he never gave the value he used. From these data, however, one can calculate this coefficient. The calculated coefficients for the above five chlorine-salinity pairs of data are: 1.80912, 1.80932, 1.80932, 1.80932, 1.80957.

These are very similar to those calculated[204] from the remaining data. Apparently the value Knudsen used was 1.809. An actual chlorine and salinity determination would not have then had accuracy greater than the third decimal. The variances in the calculated coefficients would have to be due to the fact that Knudsen's chlorine values, as presented in the data, were rounded of to the second decimal whereas in his calculations he probably multiplied by the chlorine which had been determined at least to the third decimal.

There is no indication that Knudsen ever determined this coefficient for the "Ingolf" Expedition. The value of 1.809 was too low to have come from Forchhammer and too high for Dittmar. The most likely source of this value was Otto Pettersson (143) who determined a value of 1.809, and whom Knudsen credited with being responsible for the use of the chlorine titration in modern hydrography (237, p.1).

The method that Knudsen had used for chlorine determinations at sea on the "Ingolf" was purely volumetric. This involved the use of potassium chromate as an indicator (the Mohr method). Tørnoe had suggested this method to Knudsen and helped him with it prior to the expedition (164, p.4). Once back on land after the "Ingolf" Expedition, Knudsen checked a number of these titrations at length with the Volhard method.[205] Knudsen designed a burette (pipet) specifically for these titrations to be done at sea.

TABLE XLII

A data sample from Knudsen's report of the "Ingolf" Expedition

Tid (time)		*Stations Nr.*	*Sted (position)*		*Dybke (depth)*		*Vandets Temp.*	*Vand (water)*		
Dato (date)	*Klok-keslet (hour)*	*(no. of the station)*	*Bredde (latitude)*	*Laengde Grw. (longitude)*	*Fv. (Danish fathom)*	*M. (metre)*	*(the temp. of the water)*	*Salt ‰ (salinity ‰)*	*Klor ‰ (chlorine ‰)*	*s(t/4)*
1896										
August 10	8 a.m.	137	65° 14′N.	8° 31′W	0	0	10°3	34.88	19.28	26.87
(Angust)					50	94	8°4	35.30	19.51	27.50
					100	188	7°6	35.30	19.51	27.61
					200	377	1°1	35.07	19.38	28.12
					297	559	–0°6	35.08	19.39	28.23

(164, p.82)

Probably for this reason plus his dominance as a coordinator in the entire later study on salinity, his name has become linked with this titration technique: it is generally referred to as the Knudsen method.

The care exercised throughout the entire study on salinity determination and definition was extreme. It is evident that these workers were aware of the possible impact of the study.

The water samples did not come from any particular voyage or expedition. Instead, Knudsen requested of various people and agencies to supply the needed six liter samples in bottles which he provided along with specific instructions for their use. All the bottles were filled the same way – rinsed three times, first to remove the distilled water in the bottles, and then filled only with filtered sea water sampled in the same manner and tightly corked (160, pp.17–18). The corks, for example, were treated thusly:

> Die Stöpsel wurden von der besten Sorte Kork gemacht und konisch zugeschnitten, sodass die Korkfibern senkrecht zur Konusachse standen. Die Korke wurden mehrmals in destilliertem Wasser ausgekocht, bis das Wasser sich nicht mehr durch kochen mit den Korken färbte. Nachdem die Korke dann getrocknet waren (doch nicht durch hohe Temperatur gänzlich gedarrt), waren sie fertig zum Gebrauche.
>
> (160, p.23)

In all, there were 24 samples of sea water available to the study. These were all surface samples and came specifically from the following regions: 18 were from the Baltic and North Sea (Scotland to Iceland regions), one from the Caribbean, two from the Mediterranean, and one from the Atlantic just off of Portugal. These samples were hardly exhaustive, but there was the attempt to run the gamut from high to low salinities although most of the samples were of lower salinity than the open ocean.

The scientific work of this study was divided into four segments: (*1*) the gathering and preserving of the water samples; and (*2*) the determination of the specific weights of these samples. These were handled by Knudsen. The determination of the chlorine and salt content (*3*) was handled by S.P.L. Sorensen, who at that time was the chairman of the chemistry department of the Carlsberg Laboratory,[206] at the request in August, 1899, of Knudsen. The German scientist Carl Forch was assigned to (*4*) the problem of the volume expansion of sea water. The Commission resolved that the work was to be done as soon as possible.

The results of the study can be divided into four segments: (*1*) the determination by titration of chlorine and its definition: (*2*) the gravimetric definition of salinity; (*3*) the equation relating chlorine and salinity ($S ‰ = 1.805\ Cl ‰ + 0.030$); (*4*) the tables based on this equation.

The method used by the Swedish chemist S.P.L. Sorensen with the assistance of graduate students to determine the chlorine content was in principle identical to the Mohr method, but the reaction flask (similar to a modern Erlenmeyer) was set on a scale and all measurements (after additions) were weighed. The supposed advantage was the speed of the volumetric with the accuracy of the gravimetric – and one checked the other, plus the fact that Dittmar had done this gravimetrically.

It apparently was taken for granted that the volumetric potassium chromate method (Mohr) would be used at sea to determine chlorine, but there was no further attempt to determine chlorine by this means in this study. The definition for chlorine officially accepted was:

> Unter Chlormenge versteht man deshalb das Gewicht einer mit der in 1 Kilo Meerwasser befindlichen, gesammten Halogenmenge äquivalenten Chlormenge.[207] (277, p.98)

Since various workers had arrived at different coefficients it was necessary to determine the salinity of these 24 water samples in order to accurately relate the coefficients. This Sorensen then did.

The method Sorensen chose to determine salinity was based on that of Tørnoe (291). The reason was due simply to the fact that the method Tørnoe had used was the only earlier one that gave fairly consistent results. The method involved the evaporation of the sea water to dryness and the subsequent heating of the residue to expel the last traces of water. Corrections were made for the chlorine that was lost as HCl.[208] Since he knew that chlorine was lost at elevated temperatures as HCl, he had to know the original chlorine content of the water sample so as to be able to correct the determined salinity accordingly. Sorensen's analytical checks for this were quite painstaking, and lengthy. The following paragraph will serve as a good summary of this correction problem.

> Die Meerwasserprobe, deren Chlorgehalt bekannt sein muss, wird nach Zusatz von Salzsäure und Chlorwasser eingedampft; dadurch wird alles Brom und die Kohlensäure ausgetrieben, und die organischen Stoffe werden oxydirt; nach Eindampfen zur Trockne und nach Erhitzen bis alles Wasser ausgetrieben ist, wird der Rückstand gewogen und danach in schwacher Salpetersäure gelöst, worauf die Chlormenge titrirt wird. Aus letzterer Chlorbestimmung, mit der Chlormenge der vorliegenden Meerwasserprobe vor dem Eindampfen verglichen, sieht man gleich, dass beim Erhitzen eine beträchtliche Menge von Chlorwasserstoff weggegangen ist, und die zum gefundenen Salzgewicht zu addirende Korrektion ist dann einfach die Differenz zwischen den beiden Chlorbestimmungen multiplizirt mit
>
> $$\frac{Cl_2 \div O}{Cl_2} = \text{ca. } (1 - \tfrac{16}{17}) = \text{ca. } \tfrac{55}{71}$$
>
> (277, pp.116–117)

TABLE XLIII

The gravimetric determination of salinity and chlorinity of nine sea water samples by Sorensen

Nummer des angewandten Meerwassers	*Nummer des Versuches*	*Gewicht (im Vacuum) der abgewogenen Menge Meerwasser*	*Gewicht (im Vacuum) des daraus gewonnenen, fur Verlust an Chlorwasserstoff nicht korrigierten Salzrestes*	*1 Kilo Meerwasser (im Vacuum gewogen) enthalt deshalb eine für Verlust an Chlorwasserstoff nicht korrigierte Salzmenge A (Gr.)*	*Die im gewogenen Salzrest nach Lösun in schwacher Salpetersäure gefundene Chlormenge (im Vacuum gewogen) (Gr.)*
Nr. 32	1	829.0844	2.10059	2.53363	1.056964
Nr. 33	2	835.3895	4.18640	5.01131	2.315671
Nr. 30	3	514.3552	7.10726	13.81781	3.615957
do.	4	508.8944	7.03163	13.81746	3.581848
Nr. 9	5	158.0746	2.79728	17.69595	1.416577
Nr. 10	6	159.0648	3.47288	21.83312	1.761049
Nr. 25	7	142.6364	3.88424	27.23176	1.967813
Nr. 2	8	138.0840	4.55661	32.99882	2.312402
Nr. 3	9	147.1113	4.90101	33.31498	2.489209
Nr. 23	10	111.3923	4.21199	37.81223	2.135348

[1] Hervorhebung der letzten Zahl bedeutet, dass diese erhöht ist; [2] Korrigierter Wert des Chlorgehaltes (p.27).

The definition of salinity as Sorensen finally expressed it was:

> Unter Salzmenge versteht man dann im folgenden: Die Gewichtsmenge der in 1 Kilo Meerwasser befindlichen, gelösten festen Stoffe, mit der Beschränkung, dass man alles Brom durch eine äquivalente Menge von Chlor ersetzt, alles Karbonat in Oxyd umgebildet und alle organischen Stoffe verbrannt denkt. (277, p.116)

In the official report to the Commission it was phrased:

> Als die Salzmenge ist die Gewichtsmenge der in 1 Kg Meerwasser enthaltenen anorganischen Salze zu verstehen, wenn alles Bromid und Jodid durch eine äquivalente Menge Chlorid und alles Carbonat durch eine äquivalente Menge Oxyd ersetz wird. (163, Supplement 9)

There is no practical difference between these two definitions. Neither definition mentioned the heating time which was at 480°C 72 hours (277, p.132).

Of the 24 samples used to determine the chlorinity only nine were chosen for direct salinity determinations. The reason for this was the shortage of time available. Sample No.30 (see Table XLIII) was run twice as a check for the reproducibility of the method.

In 1 Kilo Meerwasser (im Vacuum gewogen) wurde man deshalb eine Chlormenge (im Vacuum gewogen) finden	*In 1 Kilo Meerwasser (im Vacuum gewogen) findet sich eine Chlormenge (im Vacuum gewogen)*	*Differenz*	*Korrektion*	*Die Salzmengen (im Vacuum gewogen) in 1 Kilo Meerwasser (im Vacuum gewogen)*	
			c		
a	*b*	*b–a*	$(b-a)\frac{54.906}{70.906}$	*A + c*	
(Gr.)	*(Gr.)*	*(Gr.)*	*(Gr.)*	*(Gr.)*[1]	
1.27485	1.47362	0.19877	0.15892		2.688
2.52806	2.92740	0.39934	0.30923		5.321
7.03008	8.08881	1.05873	0.81988	14.6376	14.634
7.08849	do.	1.05032	0.81331	14.6308	
8.96144	10.41027	1.44883	1.12190		18.818
11.07126	12.84221	1.77095	1.37133		23.204
13.79601	16.02314[2]	2.22713	1.72458		28.956
16.74634	19.41698[2]	2.67064	2.06801		35.067
16.92058	19.59106[2]	2.67048	2.06788		35.383
19.17833	22.23709	3.05876	2.36855		40.181

(277, p.135)

The nine pairs of the values (Table XLIII, column 8 [Cl ‰] and column 11 [S ‰]) were then used by Knudsen to establish the relationship between chlorinity and salinity (the average salinity of Sample No.30 was used). As a result, then, of these chlorinity and salinity determinations, an empirical relationship between the two was established. This was expressed by the equation: S ‰ = 1.805 Cl ‰ + 0.030.

Since the method upon which the Sorensen definition is based is an extremely difficult one to perform and is beset with a multitude of technical difficulties, Knudsen proposed that salinity be defined by this equation. From this, the *Hydrographical Tables* relating chlorinity, salinity, and density were constructed (165). This equation was, in fact, first available to the oceanographic community as part of the tables published in 1901.

It should be pointed out that neither the procedural definition of salinity of Sorensen, nor that of Knudsen's equation, correspond to the total salts in a sea water sample. The values obtained for salinities by Sorensen's method would be slightly lower than those by the chlorine one. The actual salinity of a water sample, that is, the total dissolved solids in that sample, can only be obtained as the sum of the analytical results of the determined constituents. There was no attempt to calculate the sum in the study of 1899–1901.

The Knudsen equation, as it is often referred to, differed from the mathematical relationship that Dittmar and virtually everyone else had used in that it contained a constant. Thus, *it made the Knudsen definition of salinity non-conservative for the addition or removal of pure water.* Secondly, and perhaps of more far-reaching importance, *it suggested the non-constancy of the ionic proportions upon which the equation is based.* Knudsen introduced the constant:

> to compensate for the fact that chlorinity is a poor estimate of salinity for waters that are highly diluted by land drainage which tend to be low in chlorides but high in other salts. (292, p.8)

Recall that most of Knudsen's samples were from waters that were more dilute than those of the open ocean. The constant in the equation is the result of the use of Baltic Sea waters for those water samples of low concentration.

At any rate, the resolutions of the 2nd International Conference for the Exploration of the Sea (1901) contained this definition:

> By salinity is to be understood the total weight in grammes of solid matter dissolved in 1000 grammes of water.[209] The ratios between Salinity, Density and Chlorine given in Dr. Martin Knudsen's Hydrographic tables are to be adopted; and the salinity is to be calculated by the use of these tables from the determination of chlorine or from the specific gravity. (163, p.4)

While the commission defined salinity by means of the Sorensen procedure, it called for its actual determination by chlorine measurement (or specific gravity, but the chlorinity [Cl ‰] measurement was easily in more common use) with the use of the *Tables*. It is apparent that Knudsen's equation and the *Tables* were accepted as the practical definition of salinity whereas the Sorensen definition was regarded as almost purely academic. Indeed, Sorensen's procedure is so time-consuming that only nine (actually ten: two No.30's) were ever performed.[210]

There is little question but that the resolution of the 2nd International Conference has been heeded since 1902. The salinity-chlorinity relationship has, for all practical purposes, become the accepted definition of salinity, or at the very least, synonymous with it.

Knudsen regarded his study as an extension of not only his own earlier work on and after the Ingolf voyage, but primarily of Forchhammer's as well as Dittmar's.

> Im grossen Ganzen schwankt das Verhältnis zwischen den verschiedenen im Meerwasser enthaltenen Salzen unbedeutend, was schon aus den Untersuchungen von Forchhammer hervorging. Hieraus folgt, dass die

totale Salzmenge sich aus der Chlormenge oder der Dichtigkeit durch Multiplikation mit Konstanten, dem sogenannten Chlorkoeffizienten und Dichtigkeitskoeffizienten, ermitteln lässt. (159, p.11)

Yet Knudsen never said that one got a precise value for salinity with the use of the chlorine coefficient, but rather:

Hierbei wird jedoch nur ein angenäherter Wert für die Salzmenge gefunden; denn man braucht nicht die Analysen mit besonderer Genauigkeit auszuführen, um mit Sicherheit nachzuweisen, *dass der Chlor- und der Dichtigkeitskoeffizient keine Konstanten*, sondern von der Konzentration der Wasserprobe abhängig sind.[211] (159, p.11)

But if one wished to establish tables relating chlorinity, salinity and density, then such an assumption was necessary:

In den letzten Jahren wurden deshalb beim Berechnen des totalen Salzgehaltes Koeffizienten benutzt, die mit der Konzentration variierten. Die Frage, inwiefern die Konzentration allein für die Grösse der Koeffizienten bestimmend sei, ist indessen bisher unbeantwortet geblieben, und man hat daher angenommen, dass die Koeffizienten für Wasserproben mit gleichem Salzgehalt gleich waren, wenn auch die Wasserproben aus verschiedenen Meeresgebieten entnommen waren. (159, p.11)

There was another factor to consider. It is generally not recognized that all the work done by Knudsen and his associates was performed in such a short period of time and Knudsen mentioned that some precision in the chlorine determinations had to be sacrificed due to the time (159, p.9). Sorensen regretted that the time had prevented him from developing a method of salt determination into one of greater accuracy. He did feel, however, that the determination as expressed in the lengthy definition of salinity was adequate, or to about 1/2000 of the total weight of salt determined.

Wenn Rücksicht darauf genommen wird, wie viele Fehlerquellen ein Versuch wie letzterer oben beschriebener Kontrollversuch in sich schliesst (die bei früheren Versuchen gefundene Zusammensetzung der drei angewandten Stoffe, die Abwägung der drei Stoffe, die Wägungen des Filterwägeglases vor und nach dem Eindampfen und Trocknen und endlich die Fehlerquelle, welche die Methode selbst enthält), muss man einräumen, dass die erzielte Uebereinstimmung, wenn auch nicht ausgezeichnet, so doch leidlich gut ist.

Obgleich ich sehr bedauere, dass die Zeit nicht erlaubt hat, durch andere Kontrollversuche derselben Art wie gerade beschrieben die Genauigkeit der Methode weiter festzustellen, glaube ich doch sagen zu

> dürfen, dass eine Salzbestimmung, welche auf die im folgenden Abschnitt beschriebene Weise ausgeführt wird und also kaum so viele Fehlerquellen als der gerade beschriebene Kontrollversuch enthält, ein ebenso gutes Resultat als dieser geben wird, so dass die Abweichung der gefundenen Salzmenge von der wirklichen Salzmenge, wie sie oben definiert ist, sicher auf höchstens 1/2000 des Gewichtes der ganzen Salzmenge geschätzt werden kann. (277, pp.127–128)

Even with these difficulties there were obvious advantages in the use of chlorinity to determine salinity at that time. Knudsen had enumerated some in the Ingolf reports (164, p.83). Evidently the International Commission also regarded the advantages of the chlorine titration to determine salinity, especially in terms of relative simplicity, to outweigh the deviations and therefore officially standardized this method.

Considering the variances in analytical results that were known to exist, as well as the time factor under which the study was conducted, it is doubtful that Knudsen or the Commission intended that the salinity-chlorinity expression should remain in use for the next 60 years. As early as 1896, Knudsen had indicated doubt in this method:

> It may perhaps be considered as superfluous to indicate the ‰ of chlorine, when the salinity S in proportion to this appears in the tables. If for all that I have done so, it is only because I consider it as a question of time, when another system than the present one will be adopted for the indication of the salinity of sea-water. (162, p.82)

Considering the relatively small number of samples used in the study, Knudsen also expressed some doubt as to the validity of the coefficient:

> Hierbei ist jedoch nicht zu vergessen, dass diese Schlussfolgerung durch Induktion aus verhältnismässig wenigen untersuchten Wasserproben gebildet ist, sodass die Möglichkeit nicht ausgeschlossen ist, dass sie in Zukunft, wenn mehr Proben aus anderen Gegenden einer Untersuchung unterzogen worden sind, sich als unzutreffend zeigen wird. Um zu einem solchen Resultat zu gelangen, erachte ich es jedoch für nötig, dass genauere Methoden als die von uns angewandten benutzt werden müssen, wenn es sich um Wasserproben handelt, die aus den offenen Meeren entnommen sind. (159, p.12)

In short, there is every indication that Knudsen himself believed this equation to be only a temporary measure, subject to change upon the acquisition of better data and more definitive study. It would seem that in view of just Knudsen's own comments, the sole use of this equation for so many years (until 1969) is hardly warranted. Yet the salinity-chlorinity relationship as expressed by this equation and the equation itself have been used almost exclusively to the present.

The reason probably lies in the way in which the equation was presented in the *Hydrographical Tables*. They give the equation in the "Explanation of the Tables" (165, p.1) and on each page of data relating chlorinity, salinity, and density. However, the explanation contains absolutely no reference to the origins of the equation, and certainly no mention of the difficulties and the uncertainty of points such as the constancy of proportions, the value of the coefficient itself, or reasons for the presence of the constant. The weight lent by the official approval of the International Commission of these *Tables* and the relationship contained therein has also been a primary cause for this acceptance. From the *Tables* alone, one would not know of these uncertainties; and the oceanographic community at large accepted these *Tables* and all that they entail with only little reservation until recently.

One could mention the lack of statistical methods in the study of 1899–1901, especially with regard to a small number of samples, but statistical procedures were then not well understood. Or perhaps one could dwell at length on the fact that this chlorinity-salinity relationship would not hold for a number of bodies of sea water, especially the Mediterranean, which Knudsen himself recognized (159, p.11).

It is odd that although the salinity is not necessary for the determination of the very important parameter density, this notion of salinity has become such a popular and even exaggerated one. Perhaps it is due to the desire on the part of oceanographers to refer to what seems to be an intrinsically more real idea than is just that of one constituent such as the chloride (or chlorinity). Or perhaps it is because salinity as compared to density or conductivity offers the advantage of expressing a weight ratio and is therefore independent of pressure and temperature effects. But the fact remains that the chlorinity-salinity relationship, its underlying idea of constancy of ionic proportions, and all that it entails has become a much used concept in modern oceanography and the study of the physical and chemical properties of sea water.

It is recognized that it is the chlorinity (by conversion from silver nitrate titration) that has been extensively used in deriving the density of sea water. But, while important, density and its determination is not under discussion here. The matter at hand, and one which is largely a subjective matter of emphasis, is that of salinity and the Knudsen equation.

CHAPTER 9

CONCLUSIONS AND EPILOGUE

The notion of ionic proportionality has been of necessity, continually recurring in this paper. This idea that the constituents in sea water exist in constant proportions was first clearly stated by Marcet in 1819, although the germ of the idea may be found in the works of Bergman in the late eighteenth century, and implied in other works. The concept of constancy was made popular by Maury in the mid-nineteenth century. Forchhammer using this concept of constancy quantified and strengthened it and suggested the use of the "coefficient" of chlorine to determine salinity. The work of Dittmar and later that of Knudsen, Forch, and Sorensen has been accepted as demonstrating this constancy and gave the chlorinity-salinity relationship its present form: $S‰ = 1.805\ Cl‰ + 0.030$.[212] Yet no one has ever proven that the major constituents of sea water exist in constant proportions – not Forchhammer, not Dittmar, not Knudsen, nor Marcet before them.

There has always been, in oceanographic measurement, some lack of consistency in chemical data values.[213] The collection of samples, their care and subsequent chemical treatment, especially prior to 1902, had been erratic. Dittmar, for example, used soft glass bottles for sample storage and some of his samples were over three years old when they were analyzed. There would certainly have been some changes in composition (especially with regard to calcium, silicate and phosphate and sodium).

Forchhammer's samples were all from the surface; by present standards, the accuracy of his analytical technique was poor; and he ran no analyses for sodium, and only a few for potassium. With the exception of sodium, Dittmar's analytical precision was excellent. He had a definite advantage in the water samples he had at his disposal in that the set of sea water samples he analyzed represented an all-oceans coverage. Although Dittmar's results were much more consistent than those of Forchhammer, Dittmar, too, noted definite variances that he could not altogether explain away.

Knudsen, Forch and Sorensen used only surface samples. Their geographic distribution was quite limited and the chlorinity of only 12 of their total of 24 samples were indicative of water from the open ocean. It should be emphasized that the samples contained a predominance of waters from the North and Baltic Seas which were not indicative of waters from the open ocean due to the excessive dilution by land drainage. As a consequence the derived relationship would apply to these samples and not to the oceans in general.

Although Marcet postulated the constant proportionality of constituents, his data was not truly indicative of this constancy. Forchhammer, Dittmar and Knudsen never said unequivocally that the constituents of sea water were in constant proportions. All of these men mentioned the variances they found. Forchhammer noted definite variations but he explained these away primarily as being due either to local conditions or to experimental error. Dittmar, too, mentioned the variations and he ascribed these as due to or innate in experimental procedures and possibly biological activity. Yet he was forced to the conclusion that:

> When we compare the percentages of the several components with the respective means, we frequently meet with differences which lie decidedly beyond the probable limits of the analytical errors; hence, the variations must be owing partly to natural causes. (79, p.226)

But even with these variations he felt that the use of the coefficient was valid. Knudsen never said that the constituents of sea water were actually proportionally constant. In fact, he expected variations to occur.

> Soll man Tabellen über die Beziehung zwischen Chlormenge und Dichtigkeit aufstellen, so ist eine solche Annahme fast eine Notwendigkeit; die Frage is dann nur, wie ungenau sie ist. Darüber kann man sich aus dem hier ausgeführten Konstantenbestimmung ein Urteil bilden. Man wird hierbei finden, dass die aus dem Roten Meere her rührende Wasserprobe (Nr. 23) im Verhältnis zu der ganzen Salzmenge weniger Chlor enthält als die übrigen Proben, und dieses ergibt sich auch aus der Beziehung zwischen Dichtigkeit und Chlormenge. (159, p.11)

His phraseology was couched with words equivalent to approximately, nearly, etc. Nor did Knudsen expect that this equation (S ‰ = 1.805 Cl ‰ + 0.030) would hold everywhere. He was, for example, skeptical about its use.

If the constancy of constituents is viewed from the standpoint of modern oceanographic data, then it is obvious that the ratios of concentrations of major constituents actually do vary in the oceans (50, p.83)...

> that even in the case of oceanic water, there can be slight changes in the "constant" ratios between different major subdivisions of the world ocean; and that even at a given locality in the world ocean there can be appreciable changes with depth. (185, p.88)
>
> The statement, "The relative composition of the oceans is constant," requires care in interpretation. It is primarily a statement of faith, which cannot be demonstrated from measurements. The statement, "The relative composition of the oceans is *nearly* constant," is meaning-

> less unless "nearly" can be defined in some unambiguous way. The most precise definition of the term would be a statistical statement concerning the distribution of values about a mean. The only analyses that might be suitable for this purpose are those of Dittmar, since his are the only "complete analyses" on a group of samples large enough to permit an estimate of variability. Unfortunately, however, it is difficult to separate the observed variability into that produced by Dittmar's analytical methods and that representing the variability in the oceans. Using the total of all analyses it is impossible to define with clarity what is meant by "nearly" because of a relative small number of samples and inadequate sampling distribution throughout the oceans. (56, p.viii)
>
> This principle of "constant relative proportions" is often stated in text books on oceanography as a proven fact. There is no doubt that it is a reasonable approximation, but it is no more than that. (60, p.79)
>
> In discussing the implications of variations in ionic ratios, Carpenter and Carritt have pointed out a tendency which has crept into the literature to regard the relative composition of sea water as constant, and to completely disregard the fact that variations have been reported. (63, p.125)

These last two quotations from current respected chemical oceanographers help to indicate the position that the concept of constant ionic proportions has achieved. The general impression imparted by most texts is that the Knudsen equation expresses the relationship between salinity and chlorinity, and that this equation ($S‰ = 1.805\ Cl‰ + 0.030$) is one of state, that the equation identifies salinity as it is and that the lengthy Sorensen definition is synonymous – the only difference being the relative simplicity of the titrimetric determination. In other words, those investigators generally since Dittmar and as a matter of course since Knudsen, have been content to determine salinity by ratios by measuring one of the major constituents – the chloride.

The variations in the determined proportionalities that appeared in the analytical results of the works of Forchhammer, Dittmar, and Knudsen on sea water have never been emphasized. While these analysts expressed definite reservation as to the constancy of sea water their reservations have been largely overlooked and the salinity-chlorinity relationship has become a universal concept prevalent in modern oceanography. The foundation upon which the Knudsen equation lies is the assumption of the constant proportionality of ions. While this assumption has occasionally been questioned, the fact remains that this constancy and the corresponding definition of salinity have been accepted and used by the oceanographic community since 1902.

World War I and the tensions leading up to it spoiled the spirit of friendly cooperation that characterized the early ICES and stunted the growth of oceanography for almost 20 years. With the exception of stimulating advances in electronics (especially in echo sounders, ASDIC, sonar, etc.), World War II had the same affect. The seas, with the advent of the U-boat and total war, were hardly a safe place for research vessels. With the tendency to study smaller, more local regions of the oceans, this, too, being in part due to the tremendous rise in operating costs of oceanographic vessels, plus the belief in the superiority of such study due to greater control, etc., there was no definitive study of the chemical and physical properties of the sea forthcoming that might affect the status of the salinity-chlorinity relationship. Thus, its validity was generally taken for granted.

Knudsen was attempting to arrive at some uniform picture for salinity determination. There is little evidence that he tried to disprove any of Dittmar's or Forchhammer's work, but rather the reverse. Knudsen, however, took the majority of his samples from regions that Forchhammer had not included when he stated this constancy of constituents. There is also the possibility that Knudsen, Forch and Sorensen believed that their work was in reality only a preliminary study to the question of chlorinity, salinity and density relations. The geographical areas they investigated were primarily in their home waters, which at this time were beginning to come under extensive investigation by scientists. Possibly as an extension to this study, they planned in the future to look at more distant regions, which they never did. Nor did they say this should be done.

Since this relationship was an essential part of the *Hydrographical Tables* (which presented chlorinities and their corresponding salinities as well as the densities), and since sub-surface density determination was beginning to be a very important study, it seems that this chlorinity-salinity relation was accepted as part of a package deal.

Thus, a primary reason for the acceptance of this relationship lies in the *Tables* themselves. Nowhere in this small volume is there any indication that the tables might be applicable only to certain parts of the oceans. The existence of the *Tables* themselves suggest also that these properties (chlorinity, salinity, and density) are equivalent. The use of the *Hydrographical Tables* permits the conclusion of the equivalence of the terms salinity, chlorinity, and density, which, again, came directly from a belief in the constancy of relative proportions. Dayton E. Carritt and James H. Capenter, in 1959, pointed out that:

> The uncertainty of a computed value of salinity from a measured chlorinity, using equation (*1*) [Knudsen's equation] is as much as 0.04‰, this being inherent in the composition of sea water and not the results of analytical error. (292, p.6)

Not only in this uncertainty (0.04‰) inferior to results obtained from present day conductivity measurements as well as those of density it is also inferior to the innate accuracy of the chlorinity titration.

Indeed, the determination of salinity by chlorinity (chloride measurement) presupposes this constancy of relative proportions of ions, an hypothesis which is one of the foundations of physical and chemical oceanography.[214] Salinity has been important because of its use as a definitive parameter of sea water.

When Dittmar used the term salinity he meant "total salt content." It should be emphasized, however, that neither the titrimetric definition of Knudsen nor the lengthy gravimetric definition of Sorensen gives the amount of total dissolved solids. This is a serious source of confusion. The term salinity throughout the literature has been used to indicate each of these three properties, namely, the total quantity of dissolved solids, that property as defined by Knudsen's equation, and that as defined by Sorensen. While these three have similarities, they basically are different and they cannot be used interchangeably. However, as has been recently noted,

> The main source of confusion appears to be in the interpretation given to equation (*1*) and to properties of definitions and functional relationships. Equation (*1*) can be considered to be *either* (a) the definition of S ‰ *or* (b) an empirical relationship between two sea water properties S‰ and Cl‰ each of which must be defined other than by the relationship. It cannot be both, although the two meanings coincide under the special conditions of constant ionic ratios. There are fundamentally two different points of view, each with its own limitations and attributes.
>
> If equation (*1*) is used as the definition of salinity, the notions of error, precision and accuracy of salinity have no meaning. So defined, S ‰ need not even be a property of real sea water. It is merely a number established by the indicated arithmetical operations. Essentially what is done here is to say that we will define the property S ‰ in terms of Cl ‰ (otherwise defined) by the general expression: (*2*) S ‰ = $a + b$ Cl ‰ where the choice of values for a and b are without limit. All that is needed is agreement, among those using the relationship, on the values to be given to a and b. In practice they are chosen such that computed values of S ‰ correspond as closely as possible with a *property obtained under some other definition* – the Sorensen definition.
>
> (292, p.7)

Recall that this is exactly what Knudsen did. If, however, Knudsen's equation is considered to be an empirical relationship between two separately defined properties, then error, precision, and accuracy now have some meaning. Again, there is the problem of the uncertainty of determining salinity (to 0.04‰).

There are additional difficulties involved with the use of the salinity-chlorinity relationship. First of all, it is obvious that if one uses the equation S ‰ = 1.805 Cl ‰ + 0.030 to determine the salinity of distilled water, one gets the value of S ‰ = 0.030. Granted, this is not done. The salinity of distilled water is 0.000. But the equation does not say so. Recall that Knudsen (see Chapter 8) had introduced the constant 0.030 since he had to compensate for the fact that the chlorinity for waters that are highly diluted in areas of large amounts of run-off gives a poor estimate of salinity. The problem here is that salinity as defined by this equation is made non-conservative for the addition or removal of pure water. This is undesirable since it implies uncertainty of ionic proportions, and the salinity–chlorinity relationship is based on the assumed constant ionic ratio.

Salinity, as it is used, is a defined quantity both in terms of its lengthy gravimetric definition or by the definition as given by the above equation.[215] Salinity may or may not represent sea water as it is. Salinity does not represent the total salt content, but a defined quantity which is related to the method by which its determination is made (192, p.17). As Strickland and Parsons put it:

> This equation (S ‰ = 1.805 Cl ‰ + 0.030) is *purely definitive* and has no universal applicability in any chemical sense. (280, p.11)

The definition of salinity as given by Knudsen's equation is less than satisfactory.

> Firstly it assumes that all sea waters of the same salinity have the same ionic ratios; secondly it similarly assumes a definite chloride/salinity ratio for heavily diluted water, which implies that all river waters must have the same composition; thirdly it makes salinity a non-conservation property. (59, p.243)

This third point is not necessarily obvious:

> The units of salinity are stated as parts per mille, or grammes per kilogramme. So it appears logical that when two waters of different salinities are mixed, the salinity of the mixed water must be calculable by simple proportion. Thus if we mix 1 kg of water of salinity 30 ‰ with 1 kg of water of salinity 40‰ we shall have 2 kg of water containing 40 + 30 = 70 g salt, and the salinity is thus 35‰. However, if we define salinity by the Knudsen relationship, this is no longer true. The chlorinity is certainly a conservative property, from its definition. If we convert back to salinity, the result is not 35‰ but 34.993‰.[216] The difference is small, but it is illogical and may at times be important.
> (59, pp.243–244)

The term partial molar volume is a measure of the rate of change of volume with respect to quantity of a constituent.

> The partial molar volumes of sea-water constituents vary over a rather wide range of values, some being positive, others negative, indicating that the volume of solution may increase on the addition of some constituents and decrease on the addition of others.
>
> Consideration of the partial molal volumes of each of the dissolved constituents in sea water permits the conclusion that chlorinity, salinity and density need not be uniquely connected as the analogous properties of solutions of a single constituent must be. (50, p.84)

It might be argued that salinity as defined by the Sorensen definition is valid. Yet this can hardly be unequivocally stated. The salinities of only nine samples of sea water have ever been published using this procedure. The results seem satisfactory, but the determination of salinity is not an easy direction determination and has always seemed to defy analytic measurements. Nine trials, all apparently run concurrently, can hardly be considered conclusive.

Even though salinity is one of the most commonly used terms in oceanography since the mid-nineteenth century, it rarely, then, has been measured directly. So common is the term salinity that most who use it either do not know or have forgotten that it is not a fundamental quantity (59, p.243). The decision of the International Conference in 1902 to accept Knudsen's definition as a practical definition of salinity has never been questioned – until the late 1950's (60, p.77).

In summary, the constancy upon which the Knudsen equation and the *Hydrographical Tables* is based has never been experimentally demonstrated. The only certain conclusion that can be gleaned from the literature is that it has never been proven that the ionic ratios of sea water are constant. It should also be clear that the chlorinity and salinity (as well as density) are not equivalent descriptive terms (50, p.84).

In short, and this is the point, *the salinity-chlorinity relationship is not and has never been chemically sound.*

Since 1902, there have been several attempts to look at the question of salinity measurement. The only direct chemical attempt using a different method of quantitative determination of salinity was done by the chemists A.A. Guntz and J. Kocher in 1952 (122). Their method involved the precipitation of the alkaline earth metals by the addition of an excess of sodium fluoride prior to evaporation. This allowed the evaporation and ignition without the loss of hydrogen chloride. In 1964, this method was further developed by the chemical oceanographers, A.W. Morris and J.P. Riley (220).

Probably the most extensive investigation since Knudsen's into the relationship between density and chlorinity was done by W. Bein, H. Hirsekorn and L. Moller, in 1935 (13). Regarded by some as a proof of the *Hydrographical Tables*, their results showed a scatter around the Knudsen formula, but the variation was, at times, considerable and in general irregular. In reality, Bein neither proved nor disproved the relationship.

Only one major change in the concept of chlorinity-salinity relationship between 1902 and 1969 has taken place. In 1940, Knudsen and Jacobsen redefined chlorinity as:

> The number giving the chlorinity in per mille of a sea-water sample is by definition identical with the number giving the mass with unit gram of Atomgewicht silber just necessary to precipitate the halogens in 0.3285234 kilogram of the sea-water sample. (150, p.8)

This definition, which has been in use until very recently, defined chlorinity as a procedure making it independent of changes in atomic weights (for silver, chlorine, and bromine). This offers an advantage as the values for atomic weights have been changed several times since 1902.

For all practical purposes, then, salinity, chlorinity, their mutal relationship, and the methods by which they are determined were the same until 1969 as they were in 1902. Many modern chemical oceanographers were less happy with the relationship as it stood. Since 1958, there has been definite discord. Durin the last ten years the concept of salinity, a chemical one, has been complicated by the adoption of physical methods for its determination (60, p.78).

Until recently, the Knudsen titration was easily the most widely used method to determine salinity. As of late, a variety of electrical means, such as conductivity and sound velocity methods, have been used with excellent success.[217]

In October of 1961, the Hydrographical Committee of the International Council for the Exploration of the Sea, meeting in Copenhagen for the 49th Statutory Meeting, received a report of a project by the National Institute of Oceanography of England, that had been jointly supported by the ICES and the Office of Oceanography of UNESCO. The project had involved the collection of sea water samples from all over the world and the subsequent investigation of their physical and chemical properties. The results indicated that the relationship between chlorinity and density that was currently accepted was inadequate (292, p.1). It was decided at this meeting that the possibility of abandoning the definition of salinity in terms of chlorinity, which had been in force since 1902, must be faced.

In closing, the Hydrographical Committee of this 49th Statutory Meeting adopted this resolution, which was passed by the Council.

The Hydrographical Committee recommends, as a consequence of the introduction of the conductivity method as a standard method for the determination of salinity, that the Council should submit the following recommendations to Unesco:

(*i*) That the ICES, the IAPO, the SCOR and any other international oceanographic bodies deemed appropriate, be requested (a) to review present knowledge of the equations of state of sea water, in particular of the properties of chlorinity, salinity, density, conductivity and refraction index, and the relationship among these properties, (b) to consider whether redefinition of any of these properties is necessary, and (c) to advice on such further investigations as may be required.

(*ii*) That the Unesco Office of Oceanography be asked to provide the funds necessary for implementation of the above recommendations.

(292, p.2)

UNESCO, in response to this, organized a Joint Panel of the Equation of State of Sea Water. The members were nominated by UNESCO, ICES, International Association of Physical Oceanographers and Scientific Committee on Oceanic Research. This panel met at UNESCO headquarters in Paris on May 23 to 25 of 1962 (292, p.2).[218] The panel adopted a total of 13 recommendations. All of these are extremely noteworthy and valuable. For purposes here, this is especially true of these four:

That as soon as practicable Copenhagen Standard Sea Water be certified in electrolytic conductivity as well as chlorinity:

That Copenhagen Standard Sea Water be recognized internationally as the primary standard for both chlorinity and conductivity measurements as soon as recommendation (*1*) has been carried out. All laboratories now preparing independent substandards are urged to compare these as a routine with the primary standard.

That when the above recommendations have been carried out, new international oceanographic tables be computed and published.

That in order that the new definition of salinity be as nearly as is possible comparable with the old, the following procedure be adopted.

(a) That the relationship between salinity and chlorinity be arbitrarily established as S ‰ = 1.80655 Cl ‰.

(b) That an empirical relationship be computed from the data of Cox et al., connecting chlorinity with σ_0.

(c) From (a) and (b), a relationship be established between salinity and σ_0. This relationship shall then be adopted as the definition of salinity.[219] (292, pp.11–12)

There are a number of reasons that make it desirable that a new definition of salinity be in as much agreement as possible with the old one. Primarily, there is the fact that a large number of chlorinity determinations now exist and it would be preferable that these be and continue to be of use.

Conductivity measurements as a means to determine the salinity of sea water have been known for at least 60 years. Knudsen had tried this method in 1900 (167). Yet it has only recently been useful with the development of reliable electronic equipment and electrochemical techniques (59, p.247). Conductivity salinometers are now capable of a salinity determination with a typical accuracy of 0.003% (198, p.74), or in other words, it can attain a precision in salinity of 0.001‰. While conductivity is a better means to measure density than chlorinity (63, p.125), it is not to be implied that it is by any means independent of the relative ratio of the ions present in sea water. Recommendation (*1*) of the Joint panel, then, was an omen of things to come.

In 1966, the recommendation to change the Knudsen equation was adopted. In October of that year, the *International Oceanographic Tables* were published jointly by UNESCO and the National Institute of Oceanography. The preparation of these tables had been supervised by the Joint Panel (see p.191). In October of 1967, the Executive Committee of SCOR and the International Association for the Physical Sciences of the Ocean (IAPSO) endorsed these tables and the new definition of salinity.

The relationship used in the preparation of the new tables was arbitrarily taken as: S ‰ = 1.80655 Cl ‰. This equation, or the relationship it expresses, is compatible with that of the Knudsen equation. For salinities of 32‰ and 38‰ the difference between the new and old expression is only 0.026‰, and at a salinity of 35‰ they yield identical results (321, p.437).

The new recommended definition of salinity is:

$$S\text{‰} = -0.08996 + 28.29720\,R_{15} + 12.80832\,R_{15}^2 - 10.67869\,R_{15}^3 + 5.98624\,R_{15}^4 - 1.32311\,R_{15}^5 \qquad (321,\ p.438)$$

The above expression was computed by the method of least squares after chlorinity had been converted to salinities using the new equation for this relationship. The conductivity ratio (R_t) is the ratio of the conductivity of a water sample to that of one having a salinity of 35.00‰ with both samples being at the same temperature (15°C for R_{15}) and at a pressure of one atmosphere (321, p.437). This relationship between salinity and conductivity ratio was based...

> on precise determination of chlorinity and R_{15} on 135 natural seawater samples, all collected within 100 m of the surface, and including samples from all oceans and the Baltic, Black, Mediterranean, and Red Seas. (321, pp.438–439).

This definition was clearly needed. The troublesome constant 0.030 which gave distilled water a salinity of 0.030 was eliminated. Most of the water that finds its way back into the ocean is river water. Granted that although a small amount of salt is lost during evaporation, real evaporation and precipitation processes involve essentially distilled water.

The new relationship between chlorinity and salinity extends the usefulness of this concept. Chlorinity and salinity, however, contrary to even this newly stated relationship, are not equivalent properties, nor in this new definition any less dependent upon the concept of constancy of composition of sea water. Furthermore:

> The extend to which they may be cannot be decided with a high degree of confidence on the limited number of measures of the properties *now* available.[220] (61, p.246)

And:

> The relationship between conductivity and density, chlorinity and salinity are not known with a precision comparable to that of measured conductivity values. (50, p.84)

Furthermore, considering the relationship between density and chlorinity in an accuracy range of 1 in 10^5-10^6, two factors, namely the effect on density by dissolved air and fluctuations in density brought about by changes in the isotopic composition of sea water (50, p.75), do have an effect and they are generally neglected.

One other point should be made about the use of salinity as defined by chlorinity. *Any studies based on the idea that these ratios are constant are then not able to note and exploit any differences* and such differences may well be an important factor in understanding the processes that occur in the oceans.

The concept of salinity-chlorinity and its underlying assumption of the constancy proportionality of ions has been both valuable and useful. In a time when vast amounts of data were beginning to be collected, there was definite need for oceanographers to agree on some standardization. This relationship contributed in large degree to this.

In North America and in Western Europe, the numbers of measurements of salinity by conductivity now far exceed those based on the chloride titration method (59, p.244). The conductivity also has the definite advantage of the ability to give a measurement in situ.

In time it is hoped that the concept of salinity-chlorinity will be dropped altogether. The availability of Copenhagen Standard Sea Water certified not only in terms of salinity and chlorinity but also in conductivity would immeasurably aid this transition. Presumably since the new recommended definition of salinity is in terms of conductivity, this certification will follow shortly. It should be emphasized that the elimination of the salinity–chlorinity relationship is not recommended just because the conductivity measurements can be made with more precision and accuracy as well as with greater speed than the chlorinity titration. The elimination of the salinity--chlorinity relationship is recommended since the ratios of the ionic constituents in sea water are not constant – the inherent properties of the sea water itself dictate this change. The definition of salinity in terms of electrical conductivity avoids the concept of constant proportions implied in the original equation. Since the original Knudsen equation (S ‰ = 1.805 Cl ‰ + 0.030) had such a long life span, there is the possible danger that the new equation relating salinity to chlorinity (S ‰ = 1.80655 Cl ‰) might remain in active use much longer than desired. It should be viewed largely as a provisional measure.

Never again, in all probability, will salinity itself ever be measured gravimetrically. Most oceanographers now do not consider it a worthwhile determination since it is believed that the time involved to perform such a procedure or to work out a new one would produce no new knowledge sufficient enough to warrant the expenditure of time. The Knudsen-Mohr titration will continue to have some place and be used, probably as a check against conductivity measurements and especially in estuarine research since it does not require expensive equipment or facilities (198, p.74).

The concept of salinity–chlorinity has had a fairly long life span. The relationship came into being a little over 100 years ago and now seems to be clearly in the process of being abandoned. Salinity as a chemical concept may be soon be a definition which will never again be used.

There is no intent here to criticize Knudsen, Forch and Sorensen and their work. These men were asked to arrive at some measures of standardization and this they did. Hampered somewhat by a lack of time, they developed a gravimetric and volumetric definition of salinity and a relationhip between the two, as well as the *Hydrographical Tables* themselves. The trouble lay not so much in the work of these men, but in its interpretation and use by others. This concept has occupied a unique position in science since 1902. It is a concept yet it was never referred to as such. It never was called a law, yet it achieved a stature especially through usage as something more than a theory. There has never been any overwhelming analytical support for this concept – yet it was taken and used as being iron-clad.

Until recently, this theory, and again it was something more than that – almost, if not quite, a scientific fact, was not seriously questioned by the scientific community at large – and it received only token resistence from the oceanographic world. Only in the last few years when the oceanographic community began to encounter and note difficulties brought on by better and more widespread data did this herald the casting of a critical eve upon this entire relationship. It is important, however, to recall that in 1902 the oceanographic community was in definite need of basic definitions and standardization in methods, especially in the realm of salinity. The Knudsen-Sorensen definitions were a solution to the confusion that reigned. Furthermore, nothing else was available.

What is important also is that all of the scientific community accepted it. Salinity then became a defined concept. That it did or did not actually describe the salt content in sea water was unimportant. It was necessary only that scientists agree on a model and normal science could then proceed. Only when chaos appeared on the scene, as the accepted paradigm no longer adequately explained all the data, was there a change to the new definition in 1969. This represented perhaps a mild revolution in oceanography.

> On behalf of the international organizations that have endorsed the new salinity definition and the associated tables, we would like to encourage their use by all oceanographers. (321, p.438)

APPENDICES

APPENDIX I

A list of observations suggested by Robert Boyle which should be made in the compilation of a natural history for each mineral water, from his *Short Memoirs for the Natural Experimental History of Mineral Waters*, published in 1684–85 (35, vol. 4, p.794).

To the first of these three sorts of observations may be referred such heads or titles, as these.

(1) In what climate and parallel, or in what degree of latitude, the mineral water does spring up, or stagnate?

(2) Whether the spring-head, or other receptacle, do chiefly regard the east, the west, the north, or the south?

(3) Whether the water be found in a plain or valley? And if not, whether it arise in a hillock, a hill, or a mountain?

(4) And whether it be found at or near the top, the middle, or the bottom, of the rising ground?

(5) Whether the waters leave any recrement, or other unusual substance, upon the stones, or other bodies, that lie in the channels they pass through as they glide along, or the receptacles that contain them?

(6) Whether there be beneath or near the medicinal water, any subterraneal fire, that hath manifest chimneys or vents, and visibly (by night only, or also by day), burns or smokes, either constantly, or at certain periods of time?

(7) Whether at or near the mouth, or orifice, of the abovementioned chimneys or vents, there be found either flowers of brimstone, or a salt like sal-ammoniac, or some other mineral exhalations in a dry form?

(8) Whether there be under or near the course of channel of the water, any subterraneal aestuary, or latent mass, of hot, but not actually, or at least visibly, burning matters? And whether such aestuary afford an uniform heat, as to sense, or have periodical hot fits, as it were; and if so, whether these come at certain and stated times, or uncertainly or irregularly?

(9) Whether it be observed, that over the aestuary, or in some other neighbouring part of the place, where the mineral water springs, there arise any visible mineral fumes or smoak (which are wont to do it early in the morning or late in the evening,) and if such fumes ascend, how plentiful they are, or what colour, and of what smell?

(10) What is the more obvious nature of the not manifestly metalline, nor marcasitical part of the soil, which the medicinal water passes through or touches? And what are the qualities of the neighbouring soil, and the adjacent country? As whether it be rocky, stony, clayish, sandy, chalky, etc.

(11) Whether there be any ores, marcasites, or earths, (especially highly coloured ones) impregnated with mineral juices, to be met with in the course of the medicinal spring, or in the receptacle of the same water stagnant? And what these minerals are whether copperish, ferrugineous, marcasitical, etc. and whether the ores do, or do not, abound in the metalline portion? As also with what other ingredient, as spar, cauke, sulphur, orpiment, arsenick, etc. (whether innocent or hurtful) they are mingled, or else compacted together?

(12) Whether it can be discovered, that the spring of the medicinal water was common water before it came to such a place or part of the soil it runs through, and there begins to be manifestly impregnated with mineral bodies?

(13) And whether, in this case, it makes any effervescence, or other conflict, with the mineral it imbibes, or with any other water or liquor, that it meets with in its way; and whether the conflict produce any manifest heat or no?

(14) Whether, if the mineral water proposed by manifestly hot, or extraordinarily cold, the springs it flows out at, or the receptacle it stagnates in, have near it (and if, it have, how near) a spring, or well of water, of a contrary quality, as it is observed in very neighbouring springs in some few places of France, and elsewhere?

(15) Whether, when the water appears in the spring or receptacle, there appear also, either floating at the top, or lying at the bottom, or swimming between both, any drops or greater quantity of oil (like naphtha or petroleum,) or some other bituminous and inflammable substance?

(16) Whether the water be considerably altered, in quantity or quality, by the different seasons of the year, as summer, winter, etc. by the much varying temperatures of the air, as to heat, coldness, drought, etc. by the plenty, or paucity, frequency, or unfrequency, of falling rains, or snows: and what may be the bounds and measures of these alterations of the mineral water?

(35, 4, pp.799–800)

APPENDIX II

Sampling devices

Most of the water samplers in use were, at best, crude. As Marcet put it in 1819, "From all these circumstances it is easy to perceive, that the means used for raising water from great depths, have hitherto been far more uniform in their principle, or certain in their performance" (204, p.168). The first sampling device that Marcet used (see Fig. 2) was similar to one used by a Captain Phipps in his voyage to the North Pole in 1773 (201, p.164), although Marcet improved upon it greatly. The valves *V* were closed when the weight *W* struck bottom, the springs *SS* kept these valves closed. This spring innovation appears to have been Marcet's own invention. The fact that this device would sample only bottom waters was of concern to Marcet, especially in deep waters where the bottom might not be reached. He ordered built to his own specifications a superlative bottle that could be closed at any depth by the dropping of a weight along a cord (Fig.3) (201, p.166). This is the first example of a "messenger" and it was the forerunner of modern water-sampling bottles.

In addition to these two instruments, Marcet used a device designed by Sir Humphry Davy (1778–1829) to take water samples:

> The principle of this instrument may be stated in a few words. It consists in a strong copper bottle of an oblong shape, closed at its neck by a stop-cock. To this bottle is attached laterally, and in a parallel direction, a metallic tube closed at the top and open at the bottom, with an air-tight piston moving within the tube. As the open end of the tube therefore descends into the sea along with the bottle, the piston which closes the orifice of the tube is gradually forced upwards into it, as the machine sinks, the air within it being proportionally compressed; but when the piston has reached a certain part of the tube, it meets with a catch and opens the cock of the bottle, which of course, instantly fills with water; and there is an ingenious contrivance by which the machine may be set before hand, so as not to let in the water till a certain known degree of pressure is made by the superincumbent column. (201, p.167)

All of these devices, though ingenious, were subject to malfunction at depth due to lightness of the cylinder construction and leaky valves. From the data Marcet published (see Appendix III), the maximum depth sampled appears to have been 305 fathoms, although temperature measurements were made to as low as 756 fathoms (201, p.169).

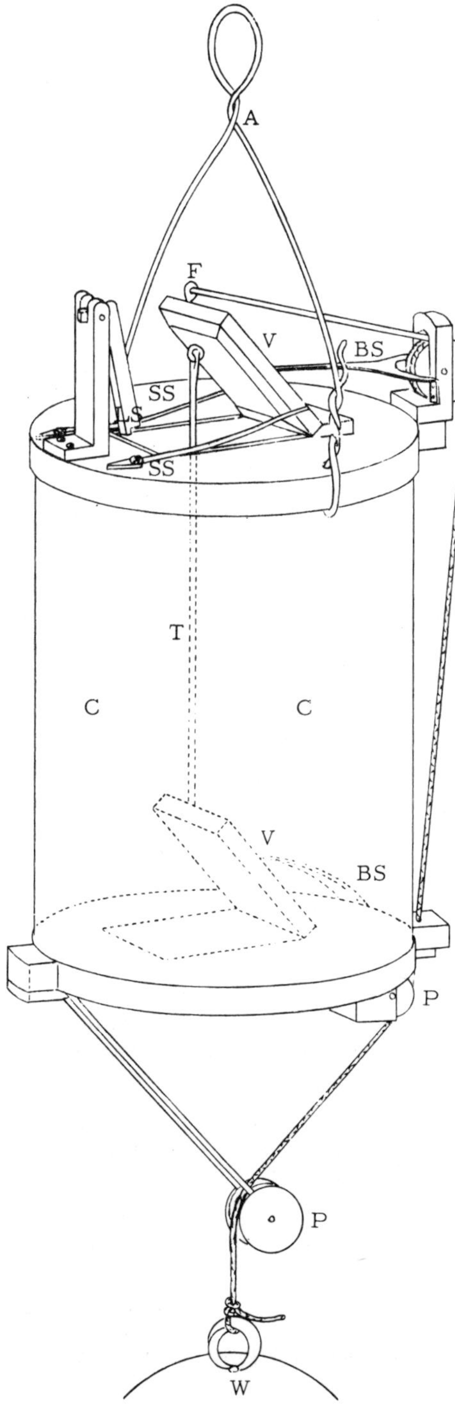

Fig. 2. Sampling bottle used by Marcet in initial studies (201, p.208).

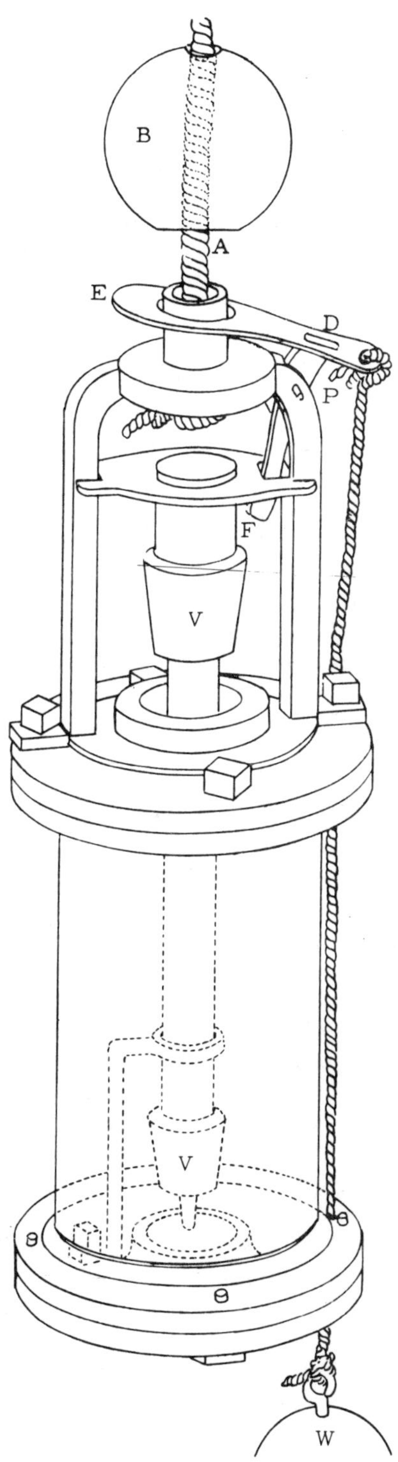

Fig. 3. Sample bottle designed by Marcet (201, p.208).

APPENDIX III

Below are the results of Alexander Marcet's investigations of the specific gravities of sea water samples taken from his "On the Specific Gravity, and Temperature, in Different Parts of the Ocean, and in Particular Seas; with some Account of their Saline Contents," published in 1819.

TABLE XLIV

Specific gravities of sea waters

Designation of seas	*No. of specimens*	*Latitude*	*Longitude*	*Specific gravity*	*Observations*
Arctic Ocean	1	66.50 N	68.30 W	1025.55	taken up by Captain Ross, in Sept. 1818, from a depth of 80 fathoms, with Sir Humphry Davy's apparatus; temperature of the water at 80 fm 30°; temperature of the air, 36°; bottle labelled in Capt. Ross's own handwriting, with all the above particulars
	2	74.0	–	1025.46*	by Lieut. Parry, from the surface, the ship surrounded by ice in every direction; temperature of the water 31°, of the air 34°, 8 July 1818
	3	74.50	59.30	1026.19	by Lieut. Parry; temperature of water 32°, of air 36°
	4	75.14	4.49 E	1027.27	by Lieut. Franklin, from the surface, 10 Sept. 1818
	5	75.14	4.49	1027.27	by Lieut. Franklin, raised with the cylindrical machine, from a depth of 56 fm; temperature of the water brought up 35°, of the air 35°, 10 Sept. 1818
	6	75.54	65.32 W	1022.7*	by Captain Ross, from the surface, 4 miles from the land, 12 August 1818
	7	75.54	65.32	1025.9	by Capt. Ross, from a depth of 80 fm with Sir H. Davy's machine; soundings 150 g, 12 August 1818
	8	76.32	76.46	1024.05*	by Capt. Ross from the surface; soundings 100 fm, 22 August 1818
	9	76.32	76.46	1026.46	by Capt. Ross from a depth of 80 fm; temperature 30.5°, 22 August 1818
	10	76.33	–	1026.64	by Lieut. Parry, with Sir H. Davy's machine, from a depth of 80 fm; temperature of the water 32° of air 36°, 21 August 1818

TABLE XLIV (continued)

Designation of seas	*No. of specimens*	*Latitude*	*Longitude*	*Specific gravity*	*Observations*
Arctic Ocean	11	79.57 N	11.15 E	1026.7	by Lieut. Franklin, from a depth of 34 fm; temperature of the sea at the surface 30.3°, at 34 fm 33.2°, of the air 35.2°
	12	80.26	10.30	1022.55*	by Lieut. Franklin, 13 July, from the surface; ship beset with ice; 12 leagues from the coast of Spitzberg; temperature of the surface 32.5°; of air 35°
	13	80.26	10.30	1027.14	by Lieut. Franklin from the bottom, depth of 237 fm.
	14	80.26	10.30	1027.15	by Lieut. Franklin from the bottom, depth of 237 fm. with Dr. Marcet's machine; temperature of the bottom 35.5°, 13 July 1818
	15	80.28	10.20	1026.8	by Lieut. Franklin from the bottom, depth of 185 fm. surface being frozen; temperature of the bottom 36½°, surface 32½°, 15 July 1818
	16	80.29	11.0	1026.84	by Lieut. Franklin from a depth of 305 fm. being the bottom; temperature of the air 36°, of the surface of the sea 32.2°, 18 July 1818

Note. The specimens marked * in the first three tables, cannot be taken into account in calculating the mean specific gravity of the waters of the ocean, their saline contents being much diminished either by the vicinity of large masses of ice, or of great rivers, which reduce them much below the average standard of density of sea water. (210, p.169)

TABLE XLV

Specific gravities of sea waters

Designation of seas	*No. of specimens*	*Latitude*	*Longitude*	*Specific gravity*	*Observations*
Northern Hemisphere	17	63.49 N	55.38 W	1026.7	by Lieut. Parry in July 1818 from a depth of 80 fm; temperature of the water 33¼°, of the surface 35°, of the air 32½°
	18	59.40	14.46	1030.04	by Capt. Basil Hall from the surface, in July 1811

TABLE XLV (continued)

Designation of seas	*No. of specimens*	*Latitude*	*Longitude*	*Specific gravity*	*Observations*
Northern Hemisphere	19	56.22 N	–	1026.56	taken up by Dr. Berger, about 15 leagues from the west coast of Jutland; depth 23 fm., Dec. 1810
	20	54.0	4.30 W	1026.8	by Dr. Berger, Calf of Man, Irish Sea
	21	53.45	0.20	1027.7	by Dr. Berger, near Hull
	22	52.45	4.0	1021.75+	by myself, from Barmouth, Wales, near the mouth of the river Mawdack
	23	48.25	6.34	1030.02	from Mr. Tennant, taken up by Mr. Lushington
	24	46.0	48.0	1026.48	by Mr. Caldewell, coast of Canada; temperature of water 42°, of air 50°
	25	45.20	45.10	1028.16	by Mr. Caldwell, brought up from a depth of 250 fm by means of a corked bottle
	26	45.10	15.0	1029.34	by Capt. Hall in January 1811
	27	25.30	32.30	1028.86	by Capt. Hall nearly in the middle of the North Atlantic
	28	22.0	89.0 E	1020.28+	by Capt. Hall from the mouth of the Ganges, about 20 miles from Calcutta; water muddy
	29	13.0	74.0	1027.72	by Capt. Hall; coast of Malabar, off Cochin; some sediment, apparently vegetable
	30	10.50	24.26 W	1028.25	by Mr. Schmidtmeyer, going to South America; bottle blackened, smell hepatic
	31	7.0	80 E	1030.9	from Mr. Tennant, by Mr. Lushington, off Colombo, Ceylon
	32	4.0	23 W	1027.72	by Mr. Schmidtmeyer in April 1808; therm. 84°
	33	3.28	81.4 E	1030.22	from Mr. Tennant, by Mr Lushington
Equator	34	0	25.30 W	1028.25	by Mr. Schmidtmeyer
	35	0	23.0	1027.85	by Capt. Hall in August 1817
	36	0	83.0 E	1028.07	by Capt. Hall in 1815, about 30 miles south of Ceylon
	37	0	92.0 E	1026.92	by Capt. Hall, 3 or 400 miles west of Sumatra, June 1817

The specific gravity of the specimens marked + in this and the following Table, being obviously much less than common, in consequence of the vicinity of rivers, these specimens have not been taken into account in calculating the mean specific gravity of sea-water. (201, p.170)

TABLE XLVI

Specific gravities of sea waters

Designation of seas	*No. of specimens*	*Latitude*	*Longitude*	*Specific gravity*	*Observations*
Southern Hemisphere	38	8.30 S	32.0 W	1028.95	taken up by Mr. Schmidtmeyer, in May 1808; temperature 82°
	39	9.0	35.0	1029.20	by Mr. Schmidtmeyer, at Pernambucco
	40	11.30	33.7	1029.80	from Mr. Tennant, by Mr. Lushington
	41	21.0	0	1028.19	by Capt. Hall, near the middle of the South Atlantic
	42	23.30	73.0 E	1028.31	by Capt. Hall, Tropic of Capricorn, between Madagascar and New Holland
	43	25.30	5.30	1032.09	by Capt. Hall, about halfway between St. Helena and the Cape in June 1815
	44	28.0	43.0	1027.15	by Capt. Hall, Mosambique, south of Madagascar
	45	35.0 #	56.0 W	1025.45	from Mr. Tennant, by Mr. Lushington, mouth of the Rio de la Plata
	46	35.10	21.0 E	1027.5	by Capt. Hall, south of the Cape, on the Banks of Lagullas
	47	35.33	0.21	1031.6	from Mr. Tennant, by Mr. Lushington, phial partly emptied
Yellow Sea	48	35.0 N#	–	1022.91	by Capt. Hall in 1816; there were several phials of this water, with glass stoppers, all the phials were blackened internally by the water, which had a highly hepatic smell; this water, when seen in large masses, has a greenish yellow colour
Mediterranean	49	36.0 N#		1030.1	by Dr. Macmichael in 1811, from a depth of 250 fm in the Straits of Gibraltar, between Cape Europe and Cabrita, with Mr. Tennant's machine
	50	36.0 N#	5.0	1030.5	by Dr. Macmichael, from the same spot as the preceding, but from the surface
	51	–	–	1027.3	by Mr. Tennant, taken up by himself at Marseilles in 1815; latitude not specified

The latitudes thus marked are stated only as approximations, not being specified on the labels of the bottles.

(201, p.171)

TABLE XLVII

Specific gravities of sea waters

Description of seas	*No. of specimens*	*Latitude*	*Longitude*	*Specific gravity*	*Observations*
Sea of Marmora	52	40.5 N	26.12 E	1028.19	taken up by Sir Robert Liston, at the entrance of the Hellespont or Dardanelles, from the bottom 34 fm deep, by my machine, in June 1812
	53	40.5	26.12	1020.28	by the same Gentleman, and exactly from the same spot as the preceeding but from the surface
	54	$41.0^{=}$	29.0	1014.44	by Sir Robert Liston, at the entrance of the Bosphorus or north entrance of the channel of Constantinople, about four miles from the land, from the bottom, 30 fm
	55	$41.0^{=}$	29.0	1013.28	by the same Gentleman; same spot, but from the surface
Black Sea	56	–	–	1014.22	by Mr. Sautter; one of the specimens clear, the other slightly hepatic, latitude not stated
	57	–	–	1014.14	same as above
White Sea	58	65.15 N	39.19 E	1018.94	by Mr. Sautter, in 1811; water perfectly clear
	59	–	–	1019.09	by the same, latitude not noted
Baltic Sea	60	56.0 N	15.0 E	1004.9	by Mr. Prevost in Carlsham harbour; cork and bottle slightly blackened
	61	57.39	–	1025.93	by Dr. Berger in 1810, Categat, one mile and a half from the eastern coast of Jutland; depth about 14 fm
Ice-Sea waters	62	56.0	12.40 E	1015.87	by Dr. Berger, from the Sound, or Passage into the Baltic, halfway between Denmark and Sweden; depth about 17 fm
	63	75.54 N	65.32 W	1000	by Capt. Ross, from the same spot as No. 6 and 7; sounding 150 fm, from an iceberg, 12 August 1818
	64	80.28	10.20 E	1000.17	by Lieut. Franklin, from water at the surface, when beset amongst ice; same spot as No. 15; temper. of the surface 32.5°, 15 July 1808
	65	79.56	11.30	1000.6	by Lieut. Franklin, from a floe, the ice being 14 ft. deep under the surface; 21 June 1818

TABLE XLVII (continued)

Description of seas	*No. of specimens*	*Latitude*	*Longitude*	*Specific gravity*	*Observations*
Ice-Sea waters	66	79.38 N	11.0 E	1000.15	by Lieut. Franklin, from an immense iceberg, August 1818
	67	76.48	13.40	1002.35	by Lieut. Franklin on 26 May 1818, about 20 miles from Spitzberg; temp. of air 29°, soundings 600 fm; taken from the surface of a small detached piece of floating ice in the sea
	68	75.40	61.20 W	1000.15	by Lieut. Parry, from young ice on the surface, about ½ inch thick, 31 July 1818

= The latitudes and longitude thus marked were inferred from the description of the spot, not being stated on the bottles. (201, p.172)

TABLE XLVIII

Presenting a synthetic view of the results obtained from the analysis of different seas; the quantity of water operated upon being in every instance supposed to be 500 grains

Description of the specimens	*Specific gravity*	*Residue (in grains) of evaporation of 500 grains of water*	*Muriate of silver*	*Sulphate of barytes*	*Oxalate of lime*	*Phosphate of magnesia*	*Total of precipitates*	*Observations*
Arctic Ocean spec. 1	1027.27	19.5	39.7	3.3	0.85	2.7	46.55	the quantity actually operated upon was 500 grains
Arctic Ocean spec. 12	1019.7	14.15	27.9	2.4	0.7	1.8	32.8	from surface; quantity operated upon 500 grs.
Arctic Ocean spec. 67	1002.35	1.75	3.2	0.1	0.05	0.03	3.37	sea ice water, coast of Spitzbergen; operated on 500 grs.
Arctic Ocean spec. 14	1027.05	19.3	38.9	3.25	0.95	2.9	46	from a depth, operated on 500 grs.
Equator spec. 35	1027.85	19.6	40.3	3.7	0.9	3.1	48	from surface, operated on 500 grs.
South Atlantic spec. 41	1028.19	20.6	40.4	3.75	1.0	3.2	48.3	operated on 250 grs.
White Sea spec. 58 & 59	1022.55	16.1	31.8	3.0	0.6	2.2	37.6	operated on 500 grs., but evaporated only 250 grs.
Black Sea spec. 56 & 57	1014.22	10.8	19.6	1.95	0.55	1.5	23.6	operated upon 500 grs. for the earths, but upon only 250 for muriate of silver and evaporation of the water
Baltic spec. 60	1004.9	3.3	7	0.7	0.2	0.6	8.5	operated upon 250 grs.; all the precipitates were slightly tinged by some vegetable or animal matter
Sea of Marmora surface, spec. 53	1020.28	14.11	28.4	2.65	0.4	2.35	33.8	entrance of Hellespont, surface; operated on 500 grs., except for muriate of silver

TABLE XLVIII (continued)

Description of the specimens	*Specific gravity*	*Residue (in grains) of evaporation of 500 grains of water*	*Muriate of silver*	*Sulphate of barytes*	*Oxalate of lime*	*Phosphate of magnesia*	*Total of precipitates*	*Observations*
Sea of Marmora bottom spec. 52	1028.19	21	40.4	3.55	0.9	3.2	48.05	from the bottom; a little carbonate of lime was deposited during evaporation; but none from the water at the surface; operated on 500 grs.
Middle of North Atlantic spec. 27	1028.86	21.3	42	3.85	0.8	2.7	49.35	operated on 250 grs. for evaporation of the water and precipitation of muriate of silver, 500 grs. for the other salts
Yellow Sea spec. 48	1022.91	16.1	32.9	1.35	0.75	2.2	37.2	during concentration deposited carbonate of lime; the water was yellowish, and had an exceedingly strong hepatic smell; proportion of magnesia rather smaller than common; operated on 500 grs.
Mediterranean spec. 51	1027.3	19.7	38.5	3.6	0.8	5.0	45.9	from Marseilles, and therefore rather weak, from the vicinity of rivers; operated on 100 grs. for evaporation and muriate of silver, and 250 for the other salts
Dead Sea	1211	195.5	325.4	0.5	9.78	55.5	584.68	Philosophical Transactions, 1807
Lake Ourmia, in Persia	1165.07	111.5	237.5	66.0	0	10.5	425.5	specimen brought by the traveller Brown; operated on 100 and 50 grs.

General Observations. In the above experiments, the residues were dried as follows, viz. The residue obtained from the water by evaporation, was thoroughly dried at a boiling heat in a water-bath, till it entirely ceased to lose weight. The muriate of silver was heated to incipient fusion, the sulphate of barytes and the oxalate of lime were dried at a boiling heat, and the ammoninco-phosphate of magnesia was heated to redness. No fiiters were used. The precipitates were washed, dried, and weighed, in the same glass capsules in which they were formed, with the exception of the magnesian salt, which was heated to redness by means of the blow-pipe, in a very thin and small platina crucible. (201, p.202)

TABLE XLIX

Showing the Differences in Temperature of Water from a Depth or Bottom, and at the Surface, observed on board His Majesty's Brig Trent, in the Arctic Seas, by Lieutenant Franklin

Date	*Latitude North*	*Longitude East*	*Depth or bottom*	*Water temperature*	*Temperature of water at surface at the time*	*Temperature of air*	*Remarks as to the situation of the vessel with respect to land or ice*
May 26	76°48′	12°26′	depth 700 fms.	43°	33°	29°	the ice in small detached pieces around the vessel; the land of Spitzbergen distant 6 or 7 leagues; the temperature of the water obtained was not tried until the bottle was taken below into the cabin, to which circumstance I think this extraordinary difference of temperature from that of the surface is to be attributed
June 20	79 58	11 25	bottom 24 fms.	31	31½	30	vessel beset by ice
June 21	79 56	–	bottom 19 fms.	31	30	30	ship surrounded by ice
June 22	80	–	bottom 33 fms.	31	30	30	surrounded by ice, not far distance from land
June 23	79 59	10 12	bottom 21 fms.	32¼	31½	30	beset in ice, close to the land
June 25	79 51	10	bottom				
			60 fms.	34	33	34	in open water, near to land;
			17 fms.	34	33	34	clear of ice, about 6 miles from land
June 26	79 44	9 33	bottom 15 fms. 34 fms.	34	34	35	in clear open water, some miles from the margin of ice; near to the land
June 27	79 51	10	bottom 72 fms.	34½	34	36	detached pieces of ice near to the vessel
June 29	79 51	10 18	bottom 17 fms. 19 fms.	34	34	39	near to the land between two islands
July 6	79 48	10 15	bottom	34½	34	36	near to the land, passing between two islands

TABLE XLIX (continued)

Date	*Latitude North*	*Longitude East*	*Depth or bottom*	*Water temperature*	*Temperature of water at surface at the time*	*Temperature of air*	*Remarks as to the situation of the vessel with respect to land or ice*
July 8	80° 20′	11° 30′	bottom 120 fms.	36	33	35	closely beset in ice – about 11 or 12 leagues from land
July 8 P.M.	80 20	11 30	bottom 130 fms.	36½	31½	33	closely beset in ice – muddy bottom
July 9 P.M.	80 26	11 38	bottom clay 120 fms. 110 fms.	36 35½	31 30½	35	beset as before – about the same distance from the nearest land
July 10	80 19	11 24	bottom 119 fms.	36	32	–	closely surrounded by ice
July 11	80 22	10 30	bottom 120 fms.	36	32	40	surrounded by ice – muddy bottom
July 12	80 20	11 7	bottom 145 fms.	35½	32	36	surrounded by ice – muddy bottom
July 13	80 22	11	bottom 217 fms.	37	32½	–	rocky bottom
July 13	80 22	11 2	bottom 235 fms.	35½	32	40½	surrounded by ice–rocky bottom
July 14	80 26	–	bottom 233 fms. muddy	35½	32	39	surrounded by ice
July 15	80 27 80 28	10 20 10 20	bottom 198 fms. 185 fms. mud	36 36¼	32 32½	38	beset amongst ice
July 16	80 26	11 26	bottom 173 fms. clay & mud	36½	32	39	closely surrounded by ice about 30 miles from land

TABLE XLIX (continued)

Date	*Latitude North*	*Longitude East*	*Depth or bottom*	*Water temper-ature*	*Temperature of water at surface at the time*	*Temper-ature of air*	*Remarks as to the situation of the vessel with respect to land or ice*
July 17	80° 27′	11°	bottom 285 fms.	35½	34	–	ice very closely besetting the vessel
July 18	80 26	10 30′	bottom 305 fms. muddy	36	32½	36	
July 19	80 24	11 14	bottom 103 fms.	36½	31½	41	the ice closely surrounding the vessel
July 20	80 21	10 12	bottom 188 fms.	35½	32½	34½	more open water than usual, distance from land 10 leagues
July 21	80 14	12 19	bottom 95 fms.	35¼	32½	41½	surrounded by ice
July 22	80 15	11	bottom 83 fms.	35¾	31	41	beset by ice
July 23	80 15	11 36	bottom 73 fms.	36¾	32½	37	the ice opening a little
July 25	80 15	11	bottom 94½ fms.	36	32½	34	the water more open than for the last fortnight
July 26	80 20	11 25	bottom 55 fms.	36	32	36	surrounded by heavy ice
Sept. 10 P.M.	75 14 75 14	3 53 3 53	depth 756 fms. 756 fms.	36 36	35 36	37	in open water, several miles distant from the margin of the ice
Sept. 24	66 35	5 33	depth 260 fms.	41½	43	44½	a bottle of this was preserved; the vessel completely in the open ocean, 300 miles from any land or ice

(201, pp.203–204)

APPENDIX IV

The silver nitrate test

The knowledge that silver nitrate would form a white precipitate with a solution of salt, including sea water, had existed since the sixteenth century (see p.40). Robert Boyle in about 1683, applied this test to the study of sea water. Count Marsilli knew of the existence and accuracy of such a test but never applied it in his experiments with the composition of sea water.

When Antoine Lavoisier and Torbern Bergman analyzed sea water 100 years after Boyle had done so, they both were familiar with the fact that the nitrate of silver could precipitate with salt in solution and they were aware of the extreme sensitivity of this test. Bergman, for example, knew that the amount of precipitate formed when silver nitrate was added to a solution of salt varied with the amount of salt in the water and that this amount could, in a clear vessel, be estimated (14, vol.1, p.172). These two men regarded this test, however, as a purely qualitative one.

Joseph Louis Gay-Lussac introduced a number of titrimetric methods of chemical analysis. Many of these appear in the many papers on volumetric analysis that he published, especially between the years 1824–1835. His volumetric methods were as a matter of policy checked in their development against gravimetric methods. To Gay-Lussac, volumetric analysis had the advantage of being simpler and faster. The most famous (283, p.270) of these methods and the one most important to the study of the sea's chemistry was his method for the analysis of silver (107). In short, it consisted of the addition of 100 ml of a sodium chloride solution of known strength to a silver solution and the settling of the precipitate. A second sodium chloride solution, 1/10 as strong as the first, was added in 1 ml amounts with subsequent shaking and settling of the precipitate. This continued until no more precipitate would occur upon addition of the salt solution. Any excess sodium chloride (Cl) was back-titrated with a dilute silver nitrate solution.

The method involved no new chemistry. C. Bartholdi (?–1849), who must be considered a pioneer of argentometry (283, p.216), had done almost the reverse in the determination of "l'acide marin" (hydrochloric acid, or chloride) in 1798 with a silver nitrate solution. Presumably, the reaction was considered complete with the cessation of precipitation (a clear-point indication). Bartholdi's results were rather inaccurate due to the uncertainty of the end point determination. Gay-Lussac overcame this by the separate addition of a dilute solution to the settled, clear solution.

In the study of sea water, a test for silver was not too important but the reverse of this method was. With the work of Gay-Lussac on the silver precipitation, the test for salt (chloride), used on and off since Boyle as a qualitative test, was made quantitative and reproducible.

By 1849 the silver nitrate precipitation was in common use in determining the chlorine content (chloride) of dissolved salts. The method, although it used the end point determination expressed by Gay-Lussac, was generally wholly gravimetric. This can easily be seen in the determination of chlorine by Usiglio and his attempt to measure the bromine. It was common knowledge by Usiglio's time that silver nitrate precipitation of chloride (HCl) also contained some bromide (HBr or Br) (283, p.184). The usual method for separating the two was based on the insolubility of barium chloride in alcohol, whereas barium bromide was soluble in this solvent. Usiglio found the results of this method lacking in consistency.

> On détermine ensuite, soit le poids de l'argent que renferme ce précipité, soit la perte de poids que l'on éprouve en traitant ce mélange par le chlore. En combinant cette détermination avec les poids atomiques du chlore et du brome, on arrive à doser l'un et l'autre de ces corps. J'ai mis en pratique ce procédé de la manière suivante. (294, p.94)

The method he used consisted of precipitation with an excess of silver nitrate ("nitrate d'argent") from a strongly acidified sea water sample.[221] The precipitant was filtered and removed and pure zinc added to react with the silver, usually in the presence of some added sulfuric acid. The silver remaining once weighed was a measure of the chlorine and bromine precipitated. The thoroughness of Usiglio's work was shown by his check of the amount of silver obtained with the silver nitrate originally used.

> Pour être certain qu'on a évité cette chance d'erreur, on dissout l'argent obtenu per l'acide nitrique pur, qui ne doit laisser aucun résidu insoluble. Le nitrate d'argent obtenu fournit

d'ailleurs, quand on le précipite par l'acide chlorhydrique, du chlorure d'argent pur dont le poids sert de vérification à la première détermination. (294, p.95)

The results obtained were consistent:

Vingt-cinq grammes d'eau de mer ont fourni un mélange de chlorure et de bromure dont le poids a été:

Echantillon no. 1, pris	1re operation	2,091 gr	
à 3000 m de la côte	2e operation	2,095	Moy. 2,0946 gr.
Echantillon no. 2, pris	1re operation	2,095;	
à 5000 m de la côte	2e operation	2,0945	

On déduit de ces nombres: brome 0,0108, chlore 0.5117, et pour 100 grammes d'eau de mer: brome 0.0432, chlore 2,0468 (294, p.95)

The silver nitrate test for chlorine (or chloride as it was then determined in salt, or hydrochloric acid) has always had particular advantages not the least of which being its extreme sensitivity. It is especially useful in sea water analysis since the chloride ion, which it precipitates very completely, is the most abundant ion in sea water. There is also the fact that this precipitation is little hindered by the difficulties other precipitating ions experience, such as co-precipitation. But the silver nitrate test, although inherently accurate, had always been troubled with the determination of its end point. The silver chloride precipitated is quite finely divided and takes time to settle; it is difficult to know just when the formation of silver chloride ceases and excess silver nitrate is being added. Gay-Lussac, as mentioned, improved this end point determination somewhat. From the standpoint of the modern concept of chlorinity there was one extremely important advancement in volumetric technique at this time. This was the needed indicator for the end point determination. This was provided by Mohr.

Friedrich Mohr (1806–1879) was one of the greatest analytical chemists of his or any time. He dabbled in many phases of science. He invented the cork borer, the Liebig condenser, and items referred to as Mohr's salt and the (Mohr's) pinch-clamp (183, p.702). He developed, for example, a quantitative method for the determination of dissolved oxygen.[222] He is even credited with the discovery of the law of conservation of energy (283, p.243).

Mohr's book, the *Lehrbuch der chemisch-analytischen Titrimethode"* (216) appearing in two parts in 1855 and 1856, was a prime impetus in the rapid advance of titrimetric analysis to its position as the most important branch of analytical chemistry. Mohr did not introduce many new methods, but he modified and improved many. Such was the nature of the man.

Mohr's solution to the silver titration problem was a stroke of genius. He let the silver act as its own indicator. To do this he simply added a dilute solution of potassium chromate to the solution to be titrated. This made the solution light yellow. The silver nitrate will preferentially precipitate chloride as long as some of the chloride remains. Once the chloride is used up the silver reacts with the chromate ion. Silver chromate is red. When the solution turns a light hazel-pink, all the chloride has been precipitated. This modification of the silver nitrate test for chloride was to become the most important method for chloride determination and would be integrally associated with the concept of chlorine (Cl^-) content of sea water. Approximately 200 years after Boyle first used silver nitrate solution as a test for sea salt, the method was now not only capable of extreme accuracy, but also capable for the first time of rapid, reliable results. Within a few years this method was in common usage.[223]

Another technique for the determination of chlorine was published 20 years after that by Mohr. This method involved the use of an alkali thiocyanate standard solution and could be used to determine silver or chlorine. It was named after the great German analytical chemist Jacob Volhard (1834–1910), who was in reality its second discoverer. The alkali thiocyanate was first used by Paul Charpentier (283, p.244) in 1870 (53). He published his results in obscure journals and, as such, they were never read by the majority of chemists (283, p.255). Volhard's paper (301) was both well read and received. The method was therefore named after him.

The idea of using the amount of titrated solution as indicative of the amount of precipitate formed goes back again to the work of Gay-Lussac. But there was the tendency prevalent, especially in sea water analysis, to prefer the gravimetric determination after the precipitation of silver chloride.

Thus, by 1895 three methods were in use to determine the amount of chlorine in a sea water sample: the Mohr method; the purely gravimetric precipitation of silver chloride with silver nitrate and the determination of the amount of silver as a measure of the amount of silver chloride formed therefore chlorine; and the Volhard method. The first and third were purely volumetric generally, but they did, upon occasion, lend themselves to being only an end point determination for a gravimetric procedure. The indicator used to determine reaction completion may well have been potassium chromate (Mohr), or ammonium ferric sulfate (Volhard), or a method such as that of Gay-Lussac. These gravimetric silver determinations are as accurate as any determination of chlorine can be. The methods are, like most gravimetric determinations, slow. A purely volumetric Mohr titration on the other hand is rapid. It suffers only from the arbitrary judgment of end-point color. The Volhard method, which takes a bit more time, is capable of the same degree of accuracy as the Mohr. Its end point is more clearly defined. At sea, the volumetric methods have a definite advantage due to inherent weighing difficulties associated with the gravimetric method when used on board ship.

In the large number of analyses of sea water that were performed at the end of the nineteenth century, there was no preference on the part of chemists in general to one method more than another (243). When Wilhelm Dittmar analyzed the water samples from the "Challenger" expedition, he used the Volhard method. He ran checks using the earlier Gay-Lussac method, as well as concurrently running his own minor modification of the Volhard method. This involved the addition of a considerable excess of silver nitrate and this, in the presence of a small amount of iron alum (potassium iron sulfate), back-titrated with the ammonium thiocyanate solution to the red color of the ferric thiocyanate. The methods are actually the same. Dittmar noted that if the resulting solution of the silver nitrate precipitation was allowed to settle for a day in the dark and then decanted the succeeding Volhard titration proceeded quite well[224]:

> I venture upon coining this word as a substitute for the customary "volumetric," which I could not well have employed, as I estimated my standard solutions by weight and not by volume (Gallice - Analyses à solutions titrées). (79, p.4)

This minor modification of the Volhard method was used by Dittmar because of his difficulty in obtaining a sharp end-point in the presence of the milky silver chloride precipitate. Furthermore, there was some question as to the validity of the Volhard method as a reagent for sea water. Dittmar's friend and fellow professor, Crum Brown, had informed him that high concentrations of magnesia salts destroyed the characteristic red color of ferric thiocyanate. Since sea water contained a significant amount of magnesia salts, Dittmar ran a rather extensive series of tests in order to determine the extent, if any, of chlorine vitiation by this presence of magnesia.

> For the deviations from the mean we have:
>
> | (1) | (2) | (3) | (4) |
> |---|---|---|---|
> | +0.002 | −0.002 | +0.005 | −0.005 |
>
> i.e., at most 5/50,000 = 0.0001 of most probable value. This shows that my apprehensions, as far as my sea water analyses were concerned, had no foundation. (79, p.8)

It is curious that Dittmar made no mention of the Mohr method for chlorine determination. In discussing chlorine determination he said:

> "Chlorine," in this section of the memoir, means total halogen calculated as chlorine. The determinations might have been made by means of the old-established process, or by the "titrimetric" process which was founded, many years ago, by Gay-Lussac upon the same reaction. This latter method would naturally suggest itself to every chemist as being the process for the case in hand, if it were not for that beautiful new method of silver-titration which was introduced, some years ago, by Volhard, and which, as regards elegance and ease of execution, is superior even to Gay-Lussac's. (79, p.4)

In his analyses, he used the Mohr burettes. The works of Mohr and more particularly his procedure for the determination of chloride were well known on the continent although they were somewhat overlooked in England. Yet Dittmar was a continental chemist. He was at least familiar with the method by 1882 (79, p.41), when he referred to the Norwegian method.[225] The Norwegian method for chlorine determination that Dittmar mentioned was actually that of Mohr.

> When using the burettes, one was filled with solution of silver and the other with sea water selected for examination, after which solution of silver was added to a flask in which the titration was performed till all chlorine had been precipitated, chromate of potassium serving as the index. The height of the fluid in both burettes was now read, and a few drops of sea water added to the mixture, to discolour it, after which solution of silver was again added, and the height of the fluids read as before, etc. After the height had been thus read 4 or 5 times in succession, the necessary data were obtained for computing the volume of sea water, which in each individual case corresponded to 1 cc solution of silver. (291, p.46)

Here Dittmar describes the first detailed and extensive use of the Mohr method for the determination of chlorine in sea water – and it was used at sea. But this method was not referred to as that of Mohr. A French chemist, Benjamin Roux, made the first positive use of the Mohr method in sea water analyses.[226] He also gave all the details, such as the first example of silver nitrate concentration in a sea water titration.

> Ce procédé consiste à employer une solution titrée d'azotate d'argent, préparee avec: azotate d'argent pur et sec, 15 g, 739 mill, et eau distillée, 984 g, 261 mill. Cette liqueur précipite complètement 5 g, 412 mill de sel marin, correspondant à 3 g, 282 mill de chlore.
>
> Pour opérer, on se sert de trois instruments: 1° une pipette de la capacité de 50 centimetres cubes; 2° une autre de la contenance de 10 grammes d'eau de mer; 3° une burette semblable à celle du chloromètre de M. Gay-Lussac, de 25 centimètres cubes, avec des subdivisions de 1/10 de centimètre cube. Quelques baguettes de verres pour servir d'agitateur, trois à quatre verres à expérience, de 200 à 250 grammes de capacité, et une solution de chromate neutre de potasse, obtenue avec 30 grammes de chromate et 270 grammes d'eau distillée, completent l'approvisionnement necessaire pour operer l'analyse de l'eau de mer. (263, p.419)

With the exception of the chemical work for the Norwegian North Atlantic Expedition of 1876–1878, the Mohr method was largely overlooked by those performing the major analyses prior to 1900. It was, however, the Mohr method that was used to determine chlorine almost entirely after 1900, and this has continued to the present. The titration procedure to determine chlorine by the Mohr method has become known as the Knudsen titration, although Martin Knudsen (1871–1949) did not add anything to the chemistry when he defined (164) the empirical relationship between chlorinity and salinity in 1902 (see p.145). No significant changes in the procedure have occurred in the time since Mohr. Mieczyslaw Oxner (237), in 1920, published a number of procedures which were designed to implement this procedure. Due to color difficulties some workers have substituted as an indicator fluorescein for potassium chromate.[227] This is the only change in the chemistry of the method and this has not been extensively used. There have been numerous improvements in the hardware used to perform the titrations which may have helped in the adoption of the Mohr method, but it is odd that the Mohr method was largely neglected for so many years, especially when one considers that it has been in almost exclusive use since the middle of the nineteenth century.

APPENDIX V

Forchhammer's analysis schemes

The analysis schemes used by Georg Forchhammer to determine the constituents of sea water taken from his *On the Composition of Sea-Water in the Different Parts of the Ocean*, published in 1865 (100).

The sulphuric acid:

> The determination of the sulphuric acid was likewise made with 1000 grains of sea-water, which, after addition of some few drops of nitric acid, was precipitated with nitrate of baryta. To try the exactness of the method three portions of sea-water were weighed, each of 3500 grains. (100, p.216)

The lime and magnesia:

To determine lime and magnesia 2000 grains (in the latter experiments only 1000 grains) were weighed, and mixed with so much of a solution of sal-ammoniac that pure ammonia did not produce any precipitate, then ammonia was added until the liquid had a strong smell thereof. It was now precipitated with a solution of the common phosphate of soda and ammonia, and filtered when the precipitate had collected into a granular powder. The precipitate thus obtained consists of tribasic phosphate of lime, and tribasic phosphate of magnesia and ammonia, which was washed with a weak solution of ammonia. All the filtered solution and the wash-water was evaporated in a steambath to dryness, and afterwards digested in a tolerably strong solution of pure ammonia, by which means there is further obtained a small quantity of the phosphates. The dry phosphates of lime and magnesia are heated, and if they are not completely white, they are moistened with a few drops of nitric acid, and again heated and afterwards weighed. The mass was not dissolved in muriatic acid mixed with alcohol until the whole contained 60 per cent (volume) thereof, mixed with a few drops of sulphuric acid, and allowed to stand for twelve hours, when the sulphate of lime is collected on a filter, heated and weighed. It contains, besides the sulphate of lime, silica, oxide of iron, phosphate of alumina, and sulphate of baryta and strontia, from which substances the sulphate of lime is separated by boiling it with a solution containing 10 percent. of chloride of sodium, which dissolves the sulphate of lime and leaves the other combinations undissolved. The remainder is washed, heated, and its weight deducted from that of the sulphate of lime. (100, p.216)

To find the quantity of magnesia contained in the weighed mixture of the phosphates of magnesia and lime, the lime, whose quantity has been determined, must, by calculation, be converted into tribasic phosphate of lime, and deducted from the whole quantity of phosphates; the other small quantities of different salts, which had been precipitated with the sulphate of lime, must likewise be deducted; the remainder is bibasic phosphate of magnesia, from which the pure magnesia is calculated. (100, p.217)

The potash (or potassium):

For a number of the analyses I have used the following method. The weighed sea-water was evaporated to dryness, the dry mass again dissolved in water, and the undissolved residue washed with warm water until all sulphate of lime is dissolved, and the wash-water does not contain any sulphuric acid. The remaining powder consists of the different after-named salts and oxides insoluble in water; it is generally weighed and noted under one head.

To this solution I add so much carbonate of lime that the sulphuric acid finds lime enough to combine with, and as much muriatic acid as would dissolve the lime of the carbonate. The quantity of carbonate of lime is determined in the following way. The equivalent of sulphate of baryta being 1456, and that of carbonate of lime being 625, there will be an excess of lime if I take carbonate of lime in such a quantity that its weight is one-half of the quantity of sulphate of baryta, obtained from an equal quantity of the same sea-water in a previous experiment for the determination of sulphuric acid. All is now evaporated to dryness and dissolved in alcohol of 60 per cent., which leaves the sulphate of lime and dissolves all the chlorides; so that the solution is quite free from sulphuric acid. It is now a third time evaporated with a sufficient quantity of chloride of platinum. Alcohol of 60 per cent. leaves the chloride of platinum and potassium, which might be weighed, and the quantity of chloride of potassium calculated from it; but as it is most difficult in a laboratory where there is constantly work going on to avoid the absorption of the vapours of ammonia by evaporating liquors, I prefer heating the double chloride to a dull red heat, and assisting the decompositon of the chloride of platinum by throwing small pieces of carbonate of ammonia in the crucible. When all the chloride of platinum is decomposed, the crucible is weighed, the chloride of potassium is extracted by alcohol of 60 per cent., and the remainder weighed again. This method has the advantage, that even if a small quantity of gypsum should have accompanied the double chloride, it will have no influence upon the determination of the chloride of potassium. When I do not want to determine the insoluble remainder, I evaporate the sea-water with a sufficient quantity of chloride of calcium, and thus leave out one evaporation and solution. (100, p.217)

Soda was determined by difference; no attempt was made to determine gases:

> I have never tried to ascertain the nature and quantity of the gases which occur in sea-water, because the collection of sea-water for that purpose would require quite different precautions from those which were necessary for the water intended for the analysis of its solid contents.
> (100, p.218)

APPENDIX VI

The results of the sea water analyses of Georg Forchhammer.
Taken from his *On the Composition of Sea-Water in the Different Parts of the Ocean,* published in 1865

Third Region: the northern part of the Atlantic, between the northern boundary of the second region and a line from the southwest point of Iceland to Sandwich Bay, Labrador

	Chlorine	*Sulphuric acid*	*Lime*	*Magnesia*	*All salts*	*Coefficient*
1. Lieutenant Skibsted, 1844	19.287	2.254	0.488	2.136	34.831	1.806
W. long. 3°15′, N. lat. 60°25′		(11.68)	(2.51)	(11.07)		
2. Captain Paludan, May 8, 1845	19.485	2.289	0.568	2.146	35.223	1.808
W. long. 5°19′, N. lat. 60°9½′		(11.75)	(2.92)	(11.01)		
3. Captain Gram, May 5, 1845	19.671	2.342	0.592	2.210	35.576	1.809
W. long. 7°52′, N. lat. 59°50′		(11.91)	(3.01)	(11.23)		
4. Captain Gram, 1845	19.619	2.296	0.587	1.820	35.387	1.814
W. long 7°20′, N. lat. 60°20′		(11.70)	(2.99)	(9.28)		
5. Captain Gram, May 7, 1845	19.620	2.306	0.581	2.189	35.493	1.809
W. long. 14°7′, N. lat. 60°9′		(11.75)	(2.96)	(11.16)		
6. Captain Gram, 1845	19.558	2.285	0.581	2.330	35.281	1.804
W. long. 16°32′, N. lat. 61°		(11.68)	(2.97)	(11.91)		
7. Taken by an Unknown	20.185	2.336	0.699	2.241	36.480	1.807
W. long. 20½°, N. lat. 55¾°		(11.59)	(3.31)	(11.10)		
8. Captain Gram, May 10, 1845	19.560	2.294	0.584	2.214	35.291	1.804
W. long. 20°30′, N. lat. 59°58′		(11.73)	(2.99)	(11.32)		
9. Captain Paludan, May 10, 1845	19.466	2.343	0.576	2.117	35.348	1.816
W. long. 23°3′, N. lat. 62°15′		(12.04)	(2.96)	(10.88)		
10. Captain Gram, May 15, 1845	19.545	2.330	0.583	2.190	35.397	1.811
W. long. 26°23′, N. lat. 59°50′		(11.92)	(2.98)	(11.20)		
11. Captain Gram	19.579	2.277	0.570	2.196	35.399	1.808
W. long. 26°37′, N. lat. 60°30′		(11.63)	(2.91)	(11.22)		
12. Captain Gram, September 1, 1845	19.386	2.365	0.578	2.135	34.990	1.805
W. long. 36°, N. lat. 58°58′		(12.20)	(2.98)	(11.01)		
Mean	19.581	2.310	0.528	2.160	35.391	1.808
		(11.80)	(2.97)	(11.03)		
Maximum	20.185	2.385	0.669	2.330	36.480	1.811
		(12.50)	(3.31)	(11.98)		
Minimum	19.287	2.254	0.488	1.820	34.831	1.804
		(11.59)	(2.51)	(9.28)		

(100,p.248)

Sea between lat. N 51° 1½′ and 55° 32′; and long. W 12° 6′ and 15° 59′

Magnesia	*Silica, etc.*	*Chloride of sodium*	*Sulphate of magnesia*	*Sulphate of lime*	*Chloride of potassium*	*Chloride of magnesium*	*All salts*	*Coefficient*
2.211 (11.24)	0.110	27.977	2.376	1.353	0.700	3.212	35.728	1.816
2.211 (11.18)	0.100	28.056	2.279	1.483	0.603	3.344	35.865	1.814
2.235 (11.35)	0.074	27.735	2.213	1.402	0.686	3.438	35.548	1.805
2.226 (11.30)	0.105	28.005	2.373	1.385	0.581	3.305	35.754	1.814
2.179 (11.03)	0.071	28.119	2.298	1.409	0.685	3.206	35.788	1.812
2.175 (11.06)	0.071	27.914	2.193	1.487	0.575	3.330	35.570	1.809
2.128 (10.83)	0.071	28.139	2.279	1.418	0.531	3.145	35.583	1.811
2.209 (11.18)	0.078	28.188	2.451	1.369	0.517	3.203	35.806	1.812
2.145 (10.92)	0.113	28.119	2.355	1.354	0.592	3.131	35.664	1.815
2.183 (11.24)	0.104	27.740	2.432	1.359	0.555	3.158	35.348	1.820
2.225 (11.34)	0.088	27.916	2.379	1.326	0.517	3.298	35.524	1.811
2.182 (11.08)	0.069	28.081	2.253	1.457	0.511	3.261	35.632	1.810
2.192 (11.15)	0.090	27.983	2.320	1.377	0.581	3.263	35.615	1.811
2.193 (11.14)	0.086	28.011	2.326	1.417	0.592	3.245	35.677	1.813
Water from the Red Sea, and from different depths in the Baltic								
2.685 (11.31)	0.136	33.871	2.882	1.676	0.612	3.971	43.148	1.818
0.403 (12.38)	0.027	4.474	0.329	0.322	0.089	0.678	5.919	1.818
0.441 (11.14)	0.072	5.810	0.632	0.333	0.092	0.526	7.465	1.886

(100, p.261)

Comparison of the means of all the regions of the Ocean (German Ocean, Kattegat, Baltic, Mediterranean, and Black Sea excepted)

		Chlorine	*Sulphuric acid*	*Lime*	*Magnesia*	*All salts*	*Coefficient*
I.	The Atlantic between the equator and N. lat. 30°	20.034	2.348 (11.75)	0.595 (2.98)	2.220 (11.11)	36.253	1.810
II.	The Atlantic between N. lat. 30° and a line from the north point of Scotland to Newfoundland	19.828	2.389 (12.05)	0.607 (3.07)	2.201 (11.10)	35.932	1.812
III.	The northernmost part of the Atlantic	19.581	2.310 (11.80)	0.528 (2.97)	2.160 (11.03)	35.391	1.808
IV.	The East Greenland Current	19.458	2.329 (11.97)	–	–	35.278	1.813
V.	Davis Straits and Baffin's Bay	18.379	2.208 (12.01)	0.510 (2.77)	2.064 (11.23)	33.281	1.811
XI.	The Atlantic between the equator and S. lat. 30°	20.150	2.419 (12.03)	0.586 (2.91)	2.203 (10.96)	36.553	1.814
XII.	The Atlantic between S. lat. 30° and a line from Cape Horn to the Cape of Good Hope	19.376	2.313 (11.94)	0.556 (2.87)	2.160 (11.15)	35.038	1.809
XIII.	The Ocean between Africa, Borneo, and Malacca	18.670	2.247 (12.04)	0.557 (2.98)	2.055 (11.01)	33.868	1.814
XIV.	The Ocean between the S.E. coast of Asia, the East Indian, and the Aleutic Islands	18.462	2.207 (11.95)	0.563 (3.05)	2.027 (11.98)	33.506	1.815
XV.	The Ocean between the Aleutic and the Society Islands	19.495	2.276 (11.67)	0.571 (2.93)	2.156 (11.06)	35.219	1.807
XVI.	The Patagonian cold-water current	18.804	2.215 (11.78)	0.541 (2.88)	2.076 (11.04)	39.966	1.806
XVII.	The South Polar Sea	15.748	1.834 (11.65)	0.498 (3.16)	1.731 (10.99)	28.565	1.814
Mean		18.999	2.258	0.556	2.096	34.404	1.811
Mean proportion of the most important substances in sea water, chlorine = 100			11.88	2.93	11.03		
Equivalents		429	45	16	82		

(100, p.257)

Comparison between the quantity of salt in sea-water from the surface and different depths in the North Atlantic Ocean

	Depth	Chlorine	Sulphuric acid	Lime	Magnesia	All salts	Coefficients
Dr. Rink, July 5, 1849 W. from Disco, N. lat. 69°45′	surface	18.524	2.268 (12.24)	0.530 (2.86)	2.119 (11.39)	35.595	1.814
	420 feet	18.532	–	0.542 (2.92)	2.098 (11.32)	–	–
Merchant Capt. Gram, May 20, 1845 W. long. 39°4′, N. lat. 59°45′	surface	19.306	2.310 (11.97)	0.575 (2.98)	2.119 (10.98)	35.067	1.816
	270 feet	19.364	2.337 (12.07)	0.579 (2.99)	2.186 (11.28)	34.963	1.806
Merchant Capt. Gram, May 5, 1845 W. long. 7°52′, N. lat. 59°50′	surface	19.671	2.342 (11.91)	0.592 (3.01)	2.210 (11.23)	35.576	1.809
	270 feet	19.638	2.338 (11.91)	0.598 (3.05)	2.210 (11.25)	35.462	1.806
Between Iceland and Greenland, Mean	surface	–	–	–	–	35.356	–
Ditto, Mean of eight samples from	1200 to 1800 feet	–	–	–	–	35.057	–
Captain Schulz, R.D.N., 1845 W. long 9°30′, N. lat. 47°45′	surface	19.644	2.556 (13.01)	0.589 (3.00)	2.273 (11.57)	35.925	1.829
	390 feet	19.640	2.595 (13.21)	0.623 (3.17)	2.357 (12.00)	35.925	1.829
	510 feet	19.699	2.594 (13.17)	0.628 (3.19)	2.296 (11.66)	36.033	1.829
Admiral von Dockum, Aug. 13, 1845 W. long. 54°15′, N. lat. 40°21′	surface	20.098	2.425 (12.07)	0.606 (3.02)	2.391 (11.90)	36.360	1.809
	210 to 270 feet	20.172	2.425 (12.02)	0.605 (3.00)	2.261 (11.21)	36.598	1.814
Capt. Irminger, Mar. 17, 1849 W. long. 64°, N. lat. 25°40′	surface	20.302	2.450 (12.07)	0.620 (3.05)	2.301 (11.33)	36.705	1.808
	2880 feet	20.222	2.380 (11.77)	0.581 (2.87)	2.274 (11.26)	36.485	1.804

Comparison between the quantity of salt in sea-water from the surface and different depths in the North Atlantic Ocean (continued)

	Depth	*Chlorine*	*Sulphuric acid*	*Lime*	*Magnesia*	*All salts*	*Coefficients*
Sir James Ross, July 29, 1843	2700 feet	20.238	–	–	–	–	–
W. long. 32°10′, N. lat. 20°54′	3600 feet	19.703	–	–	–	–	–
Sir James Ross, July 27, 1843	surface	20.429	–	–	–	–	–
W. long. 29°56′, N. lat. 18°16′	3600 feet	19.666	–	–	–	–	–
Sir James Ross, July 26, 1843	surface	20.186	–	–	–	–	–
W. long. 29°0′, N. lat. 16°57′	900 feet	20.029	–	–	–	–	–
	2700 feet	19.602	–	–	–	–	–
Sir James Ross, July 25, 1843	surface	20.081	–	–	–	–	–
W. long. 28°10′, N. lat. 15°38′	6360 feet	19.747	–	–	–	–	–
Sir James Ross, July 24, 1843	900 feet	19.934	–	–	–	–	–
W. long. 27°15′, N. lat. 14°18′	2700 feet	19.580	–	–	–	–	–
	3600 feet	19.705	–	–	–	–	–
Sir James Ross, July 22, 1843	surface	20.114	2.343	0.619	2.315	36.195	1.800
W. long. 25°35′, N. lat. 12°36′			(11.65)	(3.08)	(11.51)		
	1850 feet	19.517	2.271	0.598	2.128	35.170	1.802
			(11.64)	(3.06)	(10.90)		
Sir James Ross, July 11, 1843	surface	20.035	–	–	–	–	–
W. long. 25°6′, N. lat. 11°43′	3600 feet	19.855	–	–	–	–	–
	4500 feet	19.723	–	–	–	–	–
Sir James Ross, July 6, 1843	surface	20.070	–	–	–	–	–
W. long. 27°4′, N. lat. 6°35′	900 feet	19.956	–	–	–	–	–
	3600 feet	19.885	–	–	–	–	–
Sir James Ross, 1843	surface	19.757	2.303	0.584	2.333	35.737	1.809
W. long. 25°54′, N. lat. 1°10′			(11.66)	(2.96)	(11.81)		
	1800 feet	19.715	2.265	0.547	2.253	35.520	1.802
			(11.49)	(2.77)	(11.43)		
	3600 feet	19.548	2.322	0.545	2.239	35.365	1.809
			(11.88)	(2.79)	(11.45)		

(100, p.259)

APPENDIX VII

The table of contents for the results of the sea water analyses from the Challenger Expedition by William Dittmar
From the *Report on the Scientific Results of the Voyage of H.M.S. "Challenger" during the years 1837–76*, published in 1884 (79).

CONTENTS

APPENDIX VIII

The method for the determination of the total salt content of a sea water sample used by Hercules Tørnoe. (Taken from the *Reports of the Norwegian North Atlantic Expedition, 1876–78, published in 1880)*.

For determining the amount of salt, the only method formerly resorted to was, so far as I am aware, the simplest, viz. that of evaporating the water and then drying the residue at a proper temperature, which has been variously fixed by different chemists at from 150° to 180°. This method, however, has proved in several respects defective, as was indeed to be expected. According to Graham and others, sulphate of magnesia, the presence of which in sea-water can hardly admit of doubt, does not part with its last molecule of water till exposed to a temperature of more than 200° whereas, on the other hand, it is highly probable that partial decomposition of the chloride of magnesium contained in the salt takes place considerably below 200°. Even after the salts had been dried for about 20 hours in an air-bath at a temperature of 170°–180°, they were still found to contain, according to my experiments, a considerable quantity of water (about 15 mgr salt per gramme); dried at a lower temperature, the amount was somewhat greater. I also tested the salts for free magnesia, and found, in direct opposition to earlier statements, that, even when dried at 160°–170°, they invariable contained a very large amount, the quantity of magnesia to every gramme of dried salt being sufficient to neutralize more than 20 mgr HCl (once, when dried at 180°, even 40 mgr). For determining the free magnesia, the salts were dissolved in a given quantity of titrated sulphuric acid, and the fluid then retitrated with dilute soda-lye of known strength. With rosolic acid as the index, the final reaction was very decided.

In order to guard against the above-mentioned errors, the following mode of operation was adopted for determining the amount of salt in sea-water.

From 30 gr to 40 gr of sea-water were introduced into a thick porcelain crucible of known weight, furnished with a tightfitting cover, and evaporated on a water-bath. So soon as the salt was sufficiently dry, the crucible, with the cover on, was heated for about 5 minutes over one of Bunsen's gas-burners, then, cooled and weighed with its contents. The free magnesia liberated by the decomposition of the chloride of mangesium was now determined in the manner previously described, and the last factor necessary for computing the total amount of salt accordingly found.

This method certainly is so far open to objection, that small quantities of chloride of sodium, chloride of magnesium, or chloride of potassium may be volatilized during the process of heating, or some portion of the sulphate of magnesia be decomposed at the high temperature, and thus occasion a loss of sulphuric acid. The error, however, arising from this source will not exert any appreciable influence on the results, provided the crucible used for the operation be of thick porcelain, and have a tight-fitting cover. Thus, for instance, I found that 1.2 gr of a proportionate mixture of chloride of potassium and chloride of sodium, on being heated for the space of an hour and a quarter over one of Bunsen's gas-burners in the crucible I had used for my salt-determination, lost only 2 mgr in weight, or 0.14 mgr every 5 minutes. Moreover, it was manifest on determining the sulphuric acid and magnesia both in the water itself and in the heated residue, that, even in the event of the heating-process being much more protracted than is necessary to obtain salt free from the smallest trace of water, no serious error can result from the volatilization of chloride of magnesium or the decomposition of sulphate of magnesia. (291, pp.55–56)

APPENDIX IX

The recommendations of the Joint Panel of the Equation of State of Sea Water to aid in the resolution of the salinity problem, published by UNESCO in 1962 (292)

(3) That all laboratories co-operate with the fundamental investigations being undertaken at the National Institute of Oceanography (UK) by providing, upon request, sea water samples required for those investigations.

(5) That the empirical relationship between conductivity at 15°C and salinity defined as in (4) be established from the data of Cox et al., and this be accepted as the means for converting measured conductivity to salinities.

(6) That a relationship similarly be established between refractive index, n, at a temperature to be decided, and S ‰.

(7) That the statements of the relationships between the four measured quantities, σ_0, γ, n and Cl ‰ include an appropriate estimate of precision.

(8) That the experimental determination of the temperature and pressure effects on conductivity and density be carried out as soon as possible, and that the status of all work in progress on such determinations be reported promptly to the chairman of the panel.

(10) That in these new oceanographic tables density and specific volume functions shall be in units of mass and length (g and cm).

(11) That when values of salinity are reported in the literature or recorded in data libraries the method of measurement (e.g., conductivity, chlorinity) by which the values were obtained shall always be indicated.

(12) That instruments used for measuring electrolytic conductivity of sea water be so calibrated that their readings can be expressed in terms of absolute conductance.

(13) That these recommendations be communicated to ICES, IAOP, SCOR, IOC and other interested bodies of the Office of Oceanography, UNESCO (292, pp.11–12).

NOTES

[1] A number of the names of cities and towns in England indicate this early salt industry in that suffix *wich* (Ipswich, Norwich, Sandwich) meant a place where salt could be dug or bay salt found. The term salt forest if often encountered in reading of these regions. This simply meant a forest that provided wood for making the charcoal used in the preparation of the salt (284, p.109).

[2] In a number of places in the collected works (5) Aristotle mentioned and described carefully organisms such as clams, mollusks, crabs, urchins, numerous fish, dolphins, seals, to mention only a few. So precise were his descriptions that the famous naturalist Louis Agassiz was able to use them in some of his work with cat-fish.

"The Achelous is a river of continental Greece which runs into the sea opposite the island of Ithaca and just at the mouth of the gulf of Corinth. For many centuries the only notice taken of this account of the cat-fish of the Achelous river was to laugh at it. The passages in which there are references to it were thought to be spurious or to be simply erroneous. The cat-fish that are known in Europe do not look after their young in this fashion, though some can make a noise with their gill covers."

"This was the state of knowledge of the habits of cat-fish until the middle of the nineteenth century. About that time the matter was taken up by the distinguished Swiss naturalist, Louis Agassiz (1807–1873) of Harvard. Now in America there are cat-fish, though of different species to those known in Europe. Agassiz observed that the male American cat-fish look after their young just as described by Aristotle. This made him suspect that the story of Aristotle might be true for Greek cat-fish. It happened that he had cat-fish sent to him from the river Achelous (1856). These, he found, were a peculiar species, different from the other European cat-fish and from the American form in which he had observed the male guarding the young. Working twenty-two centuries after Aristotle, and in a continent unknown to that great naturalist, Agassiz therefore called his newly discovered cat-fish after Aristotle. Unfortunately his description was overlooked by naturalists. It was not until twenty years ago (1906) that *Parasilurus aristotelis* became properly known to men of science. That we are, even now, without information more modern than Aristotle as to the breeding of this creature gives some indication of the value of his work." (273, pp18-20).

[3] This characteristic of salt became a common test for it during the middle ages (193, p.78) – presumably it came from Aristotle.

[4] Salt separated from sea water has a high concentration of magnesium salts which give it a more bitter, less salty taste as well as causing it to lump due to the increased hygroscopicness.

[5] Book II of the *Meteorologica* (5, vol.3) presents Aristotle's views as to the nature of the sea and why it is salt. This was reinforced by additional statements in *Problemata*, Book XXIII (5, vol.7) and *De Plantis*, Book II (5, vol.4).

[6] Aristotle did not mention these people by name. The source of these theories and therefore the attributions are the footnotes of the editors in the Oxford edition (5) of Aristotle's works.

[7] Here there is a suggestion that Aristotle believed fresh water and salt water to be different.

[8] The "dry exhalation" is analogous to water vapor in that it is the same sort of thing produced by the action of heat (from the sun) on earth rather than on water. Just as the sun evaporates part of

the water (moist exhalation) it also "evaporates" some of the solid earth, hence the dry exhalation.

9 An interesting sidelight here is that Aristotle said (5, vol.7: 931^{b}9–18) that a ship floated deeper in harbor than in the open sea and he attributed this to the action of the depth of the water. The shallow water being a smaller quantity offers less resistance and the craft sinks deeper. Aristotle based much of his ideas on observations and probably actually had observed these ships. If so, then it probably was due to lower salt content and therefore lighter water in harbors – where river run-off is generally prevalent. Yet he made no density correlation between the egg in fresh and salt water and levels of ship floatage.

10 This example (which facts do not bear out) occurs in the later writings of others. This use of the wax container to purify sea water occurs numerous times in later writings. As late as in 1538 it appeared in the works of Jean Taisnier (b. 1509), and by Giambattista della Porta (1535–1615) in his *Magia Naturalis* (1589) (240, vol.2, pp.21–22, 42; 255, p.397).

11 It was this quality that made sea water poorer in extinguishing fires since a fatty substance is hotter, and also due to its dryness because of its proportionally lesser water content when compared to fresh water (5, vol.7: 935^{a}18–21).

12 Possibly Pliny was referring to the washing of the evaporated salt.

13 This idea was probably due to the known saltness of wood ashes.

14 It is clear that he meant soda (or what is now called sodium carbonate) since he said that it was used extensively in glass making.

15 This idea came from Aristotle, see p.3.

16 Probable conversions in English: 1 cyathus = 0.08 pints; 1 sextarius = 0.99 pints (*De Re Metallica*, 1, p.550).

Note that this mixture is by volume not by weight. There is no indication as to how this was arrived at. Taste was most probably the means. Implied here also is the idea of a saturated solution.

17 This almost certainly comes from Aristotle (see p.6).

18 This, too, does not occur (253, Book 31, pp.421–423).

19 Chapter 12: The Greek Sea (Mediterranean); Chapter 13: Sea of Nitus and Maritus and the Strait of Constantinople; Chapter 14: Sea of Bab el-Abivab and Jorjan (the Caspian Sea).

20 He attributes this observation to el-Kindi and Ahmad Bey el-Taibo-Es-Sarakhsi.

21 Presumably by heat transfer from the earth, although not said.

22 Several points might be mentioned here which though not pertinent to the treatment of salt content of the ocean may be of interest to oceanographers in general. One is the mention by Mas 'Udi of whales (86, p.264) in the Mediterranean. The other defined the cause of winds and storms. Winds and storms come partly from the bottom of the sea (the Arabs were interested in the bottom of the sea, especially with respect to sedimentation, much more so than the Greeks) and partly from the air although in some areas the wind arises wholly from an agitation of the air without any wind coming from the sea bottom. The sea winds, those from the sea bottom, blow from the land and penetrate into the sea. At certain times some blow in certain directions (86, p.270). This is among the earliest attempt to explain wind and storms, and to correlate both.

23 While there are advantages to distillation over evaporation (the time saved) these were not recognized here probably due to the relative crudeness of the two methods and the lack of clearly defined techniques.

24 To show this he used a test first given by Andrea Baccius in 1571 (281, p.154). Baccius had soaked a weighed cloth in the water in question, dried it and reweighed it.

25 The oak gall test was old, having been mentioned by Pliny. It was a reagent used to detect iron in

the possible profitable adulteration of copper sulfate (Verdigris) (75, p.43), copper acetate or carbonate with iron sulfate (283, p.7). It was among the earliest chemical tests. The gall "nuts" (or oak apple) were formed by an oak tree as a protective measure for the tree itself as the result of individual insect burrowing. The galls were cut up and placed in water. The extract solution turned black upon contact with iron. The purity of alum was also checked with this reagent.

26 Called a water-poise.

27 This paper contained an early reference to the "burning of the sea" or phosphorescence.

28 The common name, at that time, for salt – according to Boyle (35, vol.3, p.765).

29 This work was the first to treat at any length the temperature of the sea as well as topics such as the unevenness of the ocean bottom.

30 See p.4–5.

31 While it is of primary importance to show here Boyle's disagreement with supposed Aristotlian doctrine, there is some question whether Boyle ever read Aristotle. One wonders if Boyle had access to Aristotle in original form.

32 Boyle was quite careful in this regard. He attempted to cross-check these facts which he received from seafarers.

33 There is no indication who did the actual sampling. Considering his preoccupation with his health it is doubtful that Boyle would have done so.

34 Specific gravity bottles.

35 This change can easily be due to water. Boyle may have been referring to occlusion of water (and increase in weight thereupon) in salt cubes. The different hydration states of the salt formed as sea water is evaporated can also easily account for weight variances.

36 Distillation as he called it. Boyle used the term distillation and evaporation synonymously. When Boyle said "distillation" he sometimes meant "evaporation" as we would use the term – although his containers were generally not open but rather almost closed.

37 A detailed description of the different methods of salt production from spring, lake and sea waters is given in *De Re Metallica* (1, pp.545–562). The process of evaporation played a role in the production of the salt alum in purification by selective precipitation which under Papal auspices during the last half of the fifteenth century grew to be the first chemical industry on any really large scale (269, p.106).

38 Although he never really said so the reason he did not use evaporation techniques was probably in large part due to difficulty in reproducibility of results. It is commonly thought that Boyle developed or initiated the silver nitrate test for sea water having himself shown that evaporation techniques for salt content were unsatisfactory. Several recent articles in the literature (258, 281) apparently attribute these ideas to the *Observations and Experiments on the Saltness of the Sea*. There does not seem, however, to be much evidence in support of either of these assertions.

39 This was deliquesced potassium carbonate. Its use was first mentioned by Boyle in his *Experimental History of Colours* (vol.I) published in 1663. This was not, however, the first appearance of this in the literature. "Oyle of Tartar" (presumably potassium carbonate) was mentioned by Edward Jorden (156) in 1631 as a precipitating agent and apparently was in common use at that time.

40 These reagents were probably never really very pure so the precipitate was most likely due to chloride contained as an impurity.

41 Such as spirit of salt (HCl).

42 Apparently containing some copper.

43 The *Probierbüchlein* dating back probably to approximately 1510 was one of only three important books of mining technology of the sixteenth century prior to *De Re Metallica*. The actual author, date and place of publishing are unknown.

[44] By precipitation from silver solution ($AgNO_3$) by salt (or HCl).

[45] This was entitled "Concerning Fresh Water made out of Sea Water," printed at the desire of the patentees, in a tract entitled, "Salt Water Sweetened," by R. Fitzgerald. The letter simply informed Beal that Captain Fitzgerald (one of Beal's nearest relatives) had devised an invention for making sea water sweet (this was essentially a ship-board still) and that his Majesty (Charles II) had asked him (Boyle) to examine the sweetened water.

[46] Patentees refers to the people to whom a patent was granted; in this case for the procedure water desalinization.

[47] These values that Boyle gave are too small compared to modern values. Sea water is only about 1/30 salt.

[48] Otto Tachenius (?–1699) in 1669 (238, vol.2, p.294) pointed out that distilled water gave no precipitate with a solution of silver, whereas river and well waters did give a precipitate like a solution of common salt.

[49] This very probably was a primary reason in that Boyle was always conscious of his health.

[50] See note 25.

[51] See note 36.

[52] Note that he mentioned the use of drops of the silver solution as well as the fact that the reagent was in liquid form. This is significant because reference to the liquid state was often omitted in these early works. Boyle generally preferred the liquid state for reagents.

[53] Boyle did not give here nor did he often give any concentration for his solutions – but this was not unusual for the times.

[54] To verify this Boyle used other precipitating agents.

[55] Although Boyle did separate and filter the silver chloride, he never weighed the samples. There is no evidence in his work that he used the test as a measure of specific saltnesses of these waters.

[56] In Boyle's works on solution analysis there is the hint of titrimetric procedure. His "Experimental History of Colours" contains probably the first drop reaction ever described (35, vol.1, pp.743–744) and there is further mention of drop reactions in the *Observations and Experiments on the Saltness of the Sea* as well as later in the "Short Memoirs for the Natural Experimental History of Mineral Waters." Boyle did not, however, suggest a titrimetric analytical method but there is a clear indication of this in his works (see 193, p.10).

[57] This is a logical consequence of his belief that the sea was by and large equally salty from top to bottom. But Boyle never said this.

[58] Even if Boyle could have taken samples from a significant greater depth, his analytical procedures never could have determined a difference in salt content.

[59] The largest part of the early French writings appeared in the *Histoire de l'Académie Royale des Sciences*, Vol.I (1666–1686) (80).

[60] This was not saltpeter. It is by no means a foregone conclusion that Duclos related the nitre and the nitrous-like material. Possibly the nitrous material was due to the nitrous portion of common salt as Duclos did indicate that salt had such a quality (84, p.387).

[61] It was commonly believed that some "additive" itself and not the distillation process was responsible for the purification of the water. It would be a constantly recurring note for well over 100 years after Duclos. In 1707 the French pharmacist Etienne François Geoffroy (1672–1731) substituted open, wide glass vessels in place of the retorts which had been in use to this time (15, p.97). Generally, the evaporations were performed in almost completely closed containers and so might be better termed distillation. Often this really was not a true distillation and the distinction is many times not clear.

[62] Boyle was well respected by the French Royal Academy and most of his works were reviewed

shortly after he had published them, although the reviews might very often not be published until years later. The mention of Boyle's works on the sea as far as 1684, for example, occur on p.387 of the *Histoire de l'Académie Royale des Sciences* (80).

63 Now known to be epsom salt.

64 The number he calculated was at least 5280 million tons in a summer's day (124, p.368).

65 The nine rivers delivered 1827 million tons per day (124, p.369).

66 Halley often traveled to places great distances away in order to make these observations. In this case he went to St. Helena, since the climate of England was unsuited for astronomical observations. He also made what was one of the early purely scientific voyages to study the compass at different locales (127).

67 As late as 1685 (254) an article in the *Philosophical Transactions* and in varying passages of the *Histoire de l'Académie Royale des Sciences* 1660 to 1686 as well (80) claimed the above subterranean canals to be the method by which the water was supplied to rivers.

68 "A Short Account of the Cause of the Saltness of the Ocean, and of the Several Lakes that emit no Rivers; with a Proposal, by help thereof, to discover the Age of the World."

69 The term "residence time" was coined by T.F.W. Barth (12) in 1952. It refers to the mean length of time an element remains in the sea water. Although Halley believed that the salts from the rivers built up constantly in the ocean he was not able to offer any experimental proof of this assertion. He also said that it was possible that these saline particles might not reside in the seas for the long periods of time generally supposed but that there might be recycling processes. The only manner, however, by which he thought they might be removed was by evaporation with the water and this he knew from his pan-evaporation studies not to be the case. Since he still seemed to feel that recycling of the salts was possible although he was not able to show proof, he left the matter unresolved.

70 Or oceanographer, although this term as well as scientist, too, for that matter, would not be used for some time.

71 At one point in his writings he expressed some vexation at the interruption of his temperature measurements caused by an ensuing naval engagement with Barbary pirates in the course of which his thermometer was broken. In a later engagement he was captured by pirates and sold into slavery where he spent some years as a slave gardner in Algiers (51, p.186). Queen Christina ransomed him from this plight and he entered her service for several years. As an indication of his interest in things nautical, although he himself was not a naval officer, he made a detailed report to Queen Christina on counter currents (strong undercurrents) in the Bospourous.

72 Boyle had previously divided the sea into these same two regions. Marsilli was familiar with Boyle's writings on the sea and mentioned him several times (210, pp.2, 34).

73 This was a colorimetric indicator solution then in common use made from mauvre flowers and was the same color. Upon addition to sea water it turned a yellowish-green (210, p.25). Marsilli described this color as near that of the mineral Chrysolite.

74 Modern determinations find sea water to have a pH of 7–8.

75 In these distillations he made use of a sand-bath (210, p.25).

76 The hydrometers that Marsilli used were similar to modern ones in that they were essentially cylindrical weighted glass affairs. Marsilli's, however, had small, collar-like metal rings (0.5 to 20 grains) with which he added or subtracted weight so as to adjust the level (210, p.23).

77 This is possible since the salt used was generally sea salt and it could still show some composition change upon re-evaporation. The weights determined by the balance were only roughly equivalent to modern values. The hydrometer can and did give good relative values.

78 Quite possibly he was referring to Boyle's later works especially those on distillation (25), but this is conjecture. It is curious that although Marsilli knew of Boyle's work on sea water he did not make any mention of the silver nitrate test that Boyle had suggested.

[79] The time was reckoned by the sun's elevation.

[80] One livre equals approximately 453.6 g.

[81] It was a general practice in these times to add soap to mineral waters as a test (195, p.44).

[82] In sea water, for example, the various salts do settle out separately (although there is overlap) upon concentration of the solution. Since there is quite a bit of overlap in the fractions no one salt is entirely free from the other.

[83] This is, of course, an outstanding example of a relative statement. While solution analyses and chemistry in general were progressing rapidly from a modern standpoint there were hardly many clear ideas in Lavoisier's time as to solution chemistry. Knowledge of the nature of the material dissolved, the dissolving material and the entire solution process was vague at best.

[84] Lavoisier would have said "the state of the art," which was the manner he was prone to refer to chemistry.

[85] He used litmus as an indicator.

[86] Gionetti resolved the analysis into distinct components, whereas many of Lavoisier's fractions were still mixtures (195, pp.81–82).

[87] The ones that Lavoisier used (175, vol.1, p.589) were patterned after those of D.G. Fahrenheit (1686–1736) (87, p.140) but were graduated in a manner similar to those of Boyle.

[88] Prior to this analysis by Lavoisier there had been no published attempt to analyze the total salt content of a sea water sample qualitatively and quantitatively. Boyle, for example, had performed only a number of qualitatively tests on sea water such as the silver nitrate test.

[89] Presumably this was a closed container, so the word distillations might more aptly be used. The volume of sea water used was 40 livres taken at four places off of the coast at Dieppe (174, p.560) presumably all from the surface.

[90] The salt-dissolving power of alcohol was well known before Lavoisier's treatise (sodium chloride is slightly soluble in alcohol). By 1765 Macquer had published the results of a number of investigations of the solubilities of salts in alcohol (especially the chlorides, nitrates, and sulfates) (193). Boulduc by 1729 had used alcohol in an attempt to isolate and discover by precipitation the salts contained in water (24). This prior work of Macquer's was undoubtedly responsible in large part for Lavoisier's use of alcohol in mineral water analyses.

[91] Guettard gave him the Dead Sea samples (194, p.69). His friend Macquer also stimulated him toward water analyses (174, p.556) in general.

[92] These findings indicate the adequacy of their procedure and techniques as the values obtained were confirmed by Boussingault (34) in 1856 (116, p.164).

[93] It was common at that time to talk about a mineral (like magnesia) combined with an acid to form a salt. In fact, it was part of the general chemical theory.

[94] The comments given were essentially those of a M. l'abbé Chappe and put forth here by Lavoisier as a critique of certain ideas and observations contained therein.

[95] Modern values do not bear this out.

[96] This is doubtful.

[97] This subject had first been treated in a thesis defended on Christmas Eve, 1770, by P. Dubb of Westrogothia, a graduate student of Bergman's.

[98] In 1772, 29,168 bottles of mineral water were imported into Sweden. The following year 23,405 bottles were brought in. These represent for the times a considerable sum of money (15, p.32).

[99] The terms evaporation and distillation had been used during this and earlier time periods interchangeably. To a modern chemist evaporation means, in simple terms, that process in an open container and distillation in a closed vessel. While evaporation in open container was known and

used to some extent, it was more common either to use a rather closed container like a retort. Bergman wrote:

"A gentle evaporation is most proper, for by violent ebullition a portion of the ingredients is dissipated, nay, sometimes decomposed. A cover is necessary, to keep out the charcoal, dust, and embers; this cover must give exit to the vapours by an hole and should be kept shut until the issuing vapour is so far condensed as to prevent the dust from falling in." (14, p.156).

Perhaps the reason most of the early evaporation took place in essentially closed containers was not necessarily any leaning to the innate superiority of this method per se but that heating methods that prevailed gave rise to so much dust and other foreign matter that heating in "closed" containers became preferable. Or perhaps the reason may also have been the attempt to save any "spirits" that might be given off.

[100] The term precipitants, or Swedish equivalent, was often misleading. While, indeed, many of these were actual precipitation reagents, a large number simply were colorimetric indicators for acidity or basicity – which involved no precipitate formation.

[101] A complete list is given in (14), pp.121–154.

[102] Soap by then was a standard though not very precise test. It gave a rough indication of the water's hardness.

[103] One "kanna" an old Swedish unit of measure, equalled 2.617 liters (15, p.8).

[104] The entire method is too lengthy to treat here in any detail (14, pp.143–182). The entire process of analyses, and indeed all of his writing, is clearly put forth as he proceeded including any necessary description of the apparatus used.

[105] One of his first references to sea water mentioned, aside from the salt content, the organic matter (putrid extract) present especially in the surface waters generated by the life and death of the "innumerable crowd of animals which live there" (14, p.116). There had been numerous references to the smell of sea water prior to Bergman, usually attributed to fish in various states of decay. Later Bergman enlarged his comment:

For the innumerable croud of fish, insects, and vegetables, that grow, live, and perish in the water, as soon as they begin to grow putrid, swell, and rise to the surface, or at least such parts of them as are extracted by the water: on the surface these meet with a sufficient quantity of salt, and free access of air, circumstances which wonderfully promote putrefaction. This destruction is a necessary part of the economy of nature; and thus many circumstances, with joint force, contribute to this operation (14, vol.1, p.230).

[106] This is too simple an explanation, but a superlative one at the time it was suggested.

[107] Bergman credited Gionetti (14, p.172) with the discovery that spirit of wine (alcohol) would not dissolve sea salt. Gionetti never stated or claimed the actual originality of this specific point (112, p.38).

[108] Magnesium sulfate does exist in sea water. The magnesium chloride probably masked the presence of the sulfate since the chloride is present in larger amounts and both are soluble in alcohol. Bergman could have identified it but he did not.

[109] This was done by a Dr. Sparrman early in June, 1776 who supplied it to Bergman (14, vol.1, p.220).

[110] This method was not new and had existed in Boyle's time and had been discovered by mariners cooling their wine (27, vol.3, p.780). It was essentially the only method available. Apparently Boyle's and Hooke's apparatus had been entirely overlooked and these bottles were to be used into the nineteenth century (258, p.10).

[111] Some of the bottles may very well have filled in this manner. More likely, however, the water at depth "squeezed" into the bottle around (and through) the cork. Often as many bottles would be retrieved uncorked as corked. This method was hardly reliable and there was no way to tell precisely from what depth the trapped water had really come. Later investigations have shown that the variation of salt content with depth is slight and it is unlikely that, given the methods

employed by Bergman, any variations would have been detected. Thus this lack of exact knowledge of the depth from whence came the water sample was not then a serious problem.

112 In the case of a purely dry analysis the water would first be entirely evaporated and then separate tests performed.

113 Prior classifications of waters had existed. As early as 1702, Friedrich Hoffman (1660–1742), the renowned professor of Halle, had made a systematic classification of mineral waters (15, pp.109–110) into bitter, alkaline, salty, ferrous, and sulphurous.

114 Which at this time were one and the same.

115 The oxalic acid test was not the test for lime (calcium oxide), but rather the test for the calcium (+2) ion as calcium does not exist in water as the oxide. Bergman considered silver nitrate to be the reagent for salt but it was also recognized as the reagent for muriatic acid. In 1800 Barium chloride was referred to generally as the muriate of baryte. The word baryte was the result of the coinage of this term by Guyton de Morveau in 1787 (240, vol.3, p.532). Originally sulfuric acid was used to test for a variety of barium salts; later this was reversed (before Bergman) and barium chloride became a test for sulfuric acid (sulfate ion). The test at this time, and that which Bergman used, for barium and calcium was the precipitation as carbonates with potassium carbonate.

116 Klaproth appears never to have analyzed a regular sea water sample from the open sea – just that of the Dead Sea.

117 Seven parts salts per thousand (‰) at the surface as compared to 35‰ for the open ocean.

118 This is not in accord with modern fact. The fact does remain, however, that these authors did know that carbon dioxide was more soluble in colder water than in warm.

119 Four years later Vogel published a similar article in German which was a somewhat more detailed restatement of the work he and Bouillon-Lagrange had done (300).

120 Murray took his water from the Firth of Forth in southeast Scotland. Although he did not mention it, he evidently seemed to believe that there was no innate difference in the water he used and that which Lavoisier had used from the English Channel.

121 In modern chemistry the terms would be cations and anions.

122 In a mineral water this would also precipitate carbonate of barytes (carbonate ion) but in sea water just the sulfate.

123 The determination of magnesia had given Murray trouble. The solution was suggested by a close friend of his, William Wollaston (227, p.267).

124 Murray always used four pints of water to begin with, but expressed the values as the contents of one pint.

125 Such was not really the case. The Firth of Forth is an estuary of the Forth River. There was probably no place in this Firth that would give a water sample which was indicative of the open sea.

126 These are sulfate as barium sulfate, calcium as the oxalate, chloride as silver muriate and magnesium as the pyrophosphate.

127 With his friend J.J. Berzelius (1779–1848), the truly great Swedish chemist, he worked out, for example, the composition of "alcohol of sulphur" (carbon disulfide) (240, vol.3, p.707).

128 Marcet appears to have been interested in water analysis (199, p.207) prior to this.

129 The use of the word "modern" is interesting as only 30 years had elapsed since Lavoisier's analysis of the Dead Sea waters. There had been, of course, numerous advances in chemical analysis in this time period. Marcet then stated that (199, p.297) a Mr. Tennant had supplied him with the water samples through the courtesy of a Mr. Gordon via Sir Joseph Banks. Smithson Tennant (1761–1815) was an English chemist, Fellow of the Royal Society. It seems reasonable to suspect that

the water was supplied to Tennant by Banks, and Marcet, searching the literature for previous analyses, found Lavoisier's. Upon checking this in light of his chemical knowledge, he embarked on the analysis of these waters.

130 They soon became aware of one another's work.

131 Six years later, M.H. Klaproth (1743–1817) published an analysis of the Dead Sea (158) in which he disagreed with Marcet's results:
100 parts of the water contained: muriate of magnesia 24.20, muriate of lime 10.60, muriate of soda 7.80, total 42.60 (158, p.39).

Klaproth's method of analysis, although it involved some precipitation, was essentially that of Lavoisier's (evaporate to dryness, extract with alcohol) (194). Of Marcet's method, Klaproth remarked, "the analysis of Dr. Marcet is a good deal different, owing in all probability to the complicated processes and calculations which he followed" (158, p.39). As might be expected, this brought a response from Marcet (200) who found fault with the evaporation method and especially the alcohol extraction. Marcet concluded "It is clear, therefore, that if Mr. Klaproth had read the whole paper [Marcet's], he would have seen that our results, as to the sum total of the salts contained in the water of the Dead Sea, far from being incompatible, may agree perfectly, since he found in 100 parts of water 42.5 parts of salts (dried at a temperature which he does not specify), whilst my specimen of water yielded 41 parts dried at 180°" (200, p.134). Klaproth used as support for his own values their agreement with those of Lavoisier and that his methods coincided with work of Bergman on sea water. Marcet's method simply was the better.

132 Prior to Marcet there had been very few attempts to measure the density of sample taken at depth. Gay-Lussac had published a small number (16) of densities and salinities. Dr. John Davy, M.D. (1790–1868), the brother of Sir Humphrey Davy, had written a paper containing more extensive density determinations than Gay-Lussac (69). These measurements by Davy, all surface, were an extensive shipboard survey of density and by the balance. Several personages had done less extensive surveys (Marcet, for example), using hydrometers.

133 Table 5 (see Appendix III) presents the sum total of this data. Notice that the values for the previous analysis of Dead Sea water have been introduced as being indicative of a sea water. The "General Observations" at the bottom indicate the method used in analysis.

134 In a French report of Marcet's paper, published in 1820, these last values were omitted (202).

135 This manner of reporting is evidence of Marcet's belief in the soundness of this precipitation method and the reporting of such values which involved no inferences. Note that Marcet, in the years since his Dead Sea analysis, leaned much more toward this method.

136 The following statement helps to indicate Marcet's thoroughness: "The water, immediately on being raised from the sea, had been allowed to stand a sufficient time to deposit the earthy particles suspended in it, by which means it had become beautifully transparent. 100 pounds of the water produced only 3 grains of earthy sediment, in which I could discover nothing but carbonate of lime and oxide of iron. It is in this sediment, according to Rouelle, that mercury is to be found. I need hardly say that I could not detect in it the least particle of that metal" (202, p.451).

137 The nitrate test Marcet used for sal ammoniac was and is poor. The other substances are present only in very small amounts and thus were not detected by the methods used.

138 This exists in very small amounts.

139 In memory of his late friend Marcet, Wollaston wrote a paper (320) "On the water of the Mediterranean" which was an attempt to explain what happened to the large amounts of salt that entered the Mediterranean as the Eastern current through Gibraltar, a question that had bothered Marcet (p.179). Wollaston's determination of saline contents varied from Marcet's, the reason being due, as Wollaston pointed out, probably to the different temperatures the two men used to dry their samples. This article shed light neither on Marcet's comments on mercury or nitrate above nor on the specific method or variations he (Marcet) might have used in the paper of 1822 (202). Presumably the method was that expressed in his earlier paper (201).

140 The point about industrial salt manufacture was a bit ironic. Sea water had been evaporated on a large scale for the production of a number of salts for many years in Europe. Salt makers had known for centuries that the manner in which sea water was evaporated dictated in some way the proportion of salts they obtained. They had also been familiar for at least 100 years with the fact that these salts separated out of the sea water in an individual step-wise manner (65, p.173). Yet there was little evidence of this knowledge in the literature. This was a case where the above knowledge had been gained by these salt manufacturers in a purely empirical trial-and-error basis. In working with sea water evaporation they gained a fund of information over the years that outstripped the chemists' ideas at least with regard to the possibility of varying results from sea water evaporation.

141 Degrees Baumé refers to an old hydrometer scale developed by Antoine Baumé (1728–1804) still in some use in Europe. As the accompanying relation shows, the values for density that Usiglio obtained for density (by weighing) were good. Density here is in C.G.S. units.

Density	*Degrees*	*Density*	*Degrees*	*Density*	*Degrees*
1.00	0.00	1.13	16.68	1.27	30.83
1.01	1.44	1.14	17.81	1.28	31.72
1.02	2.84	1.15	18.91	1.29	32.60
1.03	4.22	1.16	20.00	1.30	33.46
1.04	5.58	1.17	21.07	1.31	34.31
1.05	6.91	1.18	22.12	1.32	35.15
1.06	8.21	1.19	23.15	1.33	35.98
1.07	9.49	1.20	24.17	1.34	36.79
1.08	10.74	1.21	25.16	1.35	37.59
1.09	11.97	1.22	26.15	1.36	38.38
1.10	13.18	1.23	27.11	1.37	39.16
1.11	14.37	1.24	28.06	1.38	39.93
1.12	15.54	1.25	29.00	1.39	40.68
		1.26	29.92	1.40	41.43

142 The concentration of sea water introduces additional problems. There is, for example, more chance of co-precipitation in the concentrated solution.

143 The order of this separation and deposition would be: calcium carbonate, calcium sulfate, sodium chloride, magnesium sulfate, carnalite ($KMgCl_3 6H_2O$), and magnesium chloride.

144 Halley did believe that the sea's saltness was constantly increasing and, negating regions of river run-off, that this increase was the same everywhere. He did not mention how exactly this would come about.

145 In conjunction with the study of the sea, Gay-Lussac invented a maximum and minimum thermometer (103).

146 Gay-Lussac collected these samples himself at altitudes of 21,460 ft. and 21,790 ft. from a balloon ascent using previously evacuated glass globes (108).

147 Due to the difficulty of operation of balances and laboratory tests on a rolling ship, these had largely given way to the more accurate method (the hydrometer) especially for precision work on board ship.

148 This does occur upon heating.

149 Such is not the case.

150 The above quote represents the first time that the element chlorine was mentioned in a paper concerning sea water. This element had been discovered by the Swedish chemist Carl Wilhelm Scheele (1742–1786) in 1774 and was isolated and shown to be an element by Sir Humphry Davy (1778–1829) in 1810 (239, pp.104, 185).

151 For all practical purposes, the French word "salure" could be translated as meaning total salt content.

152 The values for density were quite good. Note the use of the average.

153 See p.90.

154 In choosing an average modern value for salinity (S‰), the figure 35‰ is generally cited. This, of course, corresponds to Gay-Lussac's values of "trois centièmes et demi."

155 Within a year the Rev. John Fleming taking samples in the Firth (or Frith) of Tay in Scotland, on the North Sea, dealt with this same point (90). Sampling through a vertical profile, he noticed a decided layering of fresh water on top of the sea water and that the fresh water remained surprisingly unmixed and maintained its freshness for some distance at sea.

156 Such is not the case. Salinity increases slightly with depth on the average, at least for the area from which Gay-Lussac had his samples taken.

157 It appears that the entire question of salts in the sea gave rise to several later articles, such as (106).

158 From a modern oceanographic standpoint the concept of a water column is in constant use.

159 Since these solutions were of single salts (generally NaCl), evaporation to dryness was an adequate technique.

160 Humboldt had said salt wells were richer in salt nearer the bottom and d'Arcet had reported that large reservoirs of soda in solution were of greater concentration at the bottom (107, p.80).

161 The apparent problem was the size of the equipment – much longer tubes would have been needed.

162 The French word "titre" refers to the purity of noble metals. It had appeared much earlier in French analytical papers (283, p.197), yet Gay-Lussac first used the verb "to titrate" (283, p.222). Since his methods in this field became so widespread so fast, it is reasonable to assume that the term "titration" stems from this use and its associated methods. The development of titrimetry is not a question here. For one wishing to look further, the very excellent book by E. Rancke Madsen (195) is strongly recommended.

163 The densities were measured using a specific gravity bottle, and given in the form of tables (see Appendix III).

164 According to later studies this is only partially true.

165 In general, Marcet's conclusions correspond approximately with more modern thinking on the subject.

166 For distilled water the value for maximum density is closer to 39°C. This conclusion is true. There is no density maximum above its freezing point for sea water. (Table XLIX, Appendix III is a representative data table for temperature data as presented by Marcet).

167 Although the device is usually credited elsewhere, Aimé introduced reversible outflow thermometers (228, p.74).

168 The eighth edition, 1861.

169 The title for the chapter in the sixth edition of the "Sailing Directions" had been suggested to Maury by Alexander von Humboldt (179, p.68).

170 In England, where it ran to 19 editions, the book had the title "Physical Geography of the Sea and its Meteorology."

171 Note that these are given as the salts themselves.

172 This is reminiscent of Gay-Lussac or Von Humboldt.

173 The *Flying Cloud* was a beautiful thing, peculiarly American. She was 208 ft. long on the keel, a

registered tonnage of 1,783, and packed 13,000 running yards of canvas. In seaworthiness and speed she was never equalled.

[174] The bibliography of this dissertation gives some indication of the scope of his works. Aside from his interest in the chemistry of the sea he published numerous works on geology and a number dealing with organic compounds.

[175] "To explain all of the neptunian formations, an exact analysis of the sea water from different parts of the world ocean is unavoidably necessary, and the author has started on this wide reaching and tedious work."

[176] The manner by which the other acids and bases were determined is given in the Appendix V.

[177] The second column "Silica, etc." would include the silica, phosphate, oxide of iron, carbonate, in short, all insoluble materials. The column "All salts" was calculated by totaling all of the other columns.

[178] He noted also that the waters near an ice cap contained greater sulfate than is normal in sea water.

[179] Or general, as Forchhammer called it.

[180] His values differ from modern ones, however, due to subsequent changes in accepted values of atomic weights.

[181] Vincent used the term alcalinity ("alcalinité") with reference to the basic quality of sea water. Forchhammer thought that sea water was neutral, assuming that the "acids and bases" present neutralized one another.

[182] This, by the way, was directed more towards recovering the silver than a check on the volumetric procedure, evidence no doubt of a professor's pay and French frugality.

[183] The "Challenger" was primarily a wind-driven vessel. Her steam power was used only during the sounding, dredging, and trawling operations. Between the 362 separate stations she traversed by sail. The use of steam made difficult operations like dredging (not to mention the hauling in tons of wet hemp line) possible. Under sail these would have been extremely difficult if not impossible, especially for sampling deep waters. Though not generally realized, the use of steam and steam-driven vessels was an extreme impetus to the development of oceanography.

[184] *The Depth of the Sea* was the result of the "Porcupine's" voyage and was published in 1872 just as the "Challenger" expedition left England.

[185] Thomson et al. were, for example, the first to show that animal life existed at a depth of one or two miles (230, p.11).

[186] The word "oceanography" was used by Dittmar in 1883 (79, pp.22, 205) and although he is sometimes regarded as its originator (see, for example, Oceans *Encyclopedia Britannica*, 1967, 235) he did not originate the term. He did, however, invent the term "alkalinity" as applied to salt water (79, pp.22, 205). Sir John Murray in his little book *The Oceans*, published in 1910, had this to say: "The term Thalassography has been used, largely in the United States, to express the science which treats the ocean. The term Oceanography is, however, likely to prevail. The Greeks appear to have used the word Thalassa almost exclusively for the Mediterranean, whereas the almost mythical "Oceanus" of the ancients corresponds to the ocean basins of the modern geographer. In recent times I believe the word Oceanography was introduced by myself about 1880, but I find from Murray's English Dictionary that the word "océanographie" was used in French in 1584, but did not then survive (230, p.11). The German word "Ozeanographie" (now often replaced by "Meereskunde") is a few years older than the English" (230, p.11).

[187] In view of the current discussion between the words oceanography and oceanology, this comment by Murray (written in 1910) might be of interest.

The words Oceanography and Oceanology are not "mongrel" words; on the contrary, they are both absolutely correct formations, on such analogies as geography, topography, and theology, demonology, anthropology, zoology. The Greek dictionary knows such a word as Thalassographos, but not Oceanographos. But to insist on this point would be the merest pedantry, for even now it is not of the Atlantic, Pacific, and Arctic Seas that we speak, but of the oceans bearing those names. Sutherland Black says: "By Thalassography the Greek dictionary chiefly means the description of the Mediterranean. A very myopic pedant might raise some scruple over '-graphy' on the ground that a mythographer is a writer of myths, and a logographer a writer of prose; but then a topographer is not a writer of places, but a describer of them: so also with geographer" (230, p.11).

188 As to the "Challenger" herself, Buchanan had these remarks: The material collected at each station had to be examined, preserved and stored, before the ship arrived at the next one. The stations were generally about 200 miles apart, so that in the passage from one port to another a station was made every second or third day. This was easily accomplished under sail and it added enormously to the comfort and the interest of the voyage. All the advantages of having a wooden sailing ship were not fully realised at the time. It was not until I had taken part in one or two expeditions in well found iron or steel ships in tropical waters that I found out the discomfort which we escaped by being on board of an "old wooden ship." The temperature of the air in the ship was, of course, never lower than that of the air outside; but, on the other hand, it was never higher. Nothing astonished me more than the perfect uniformity of temperature of the air of the main deck of the ship in the tropics. I was able to make experiments on the effects of pressure on the deep sea thermometers in a hydraulic apparatus on the main deck, which I could not have made anywhere else. The temperature of the air did not vary by one-tenth of a degree (C.) during the whole of the day (42, p.36).
and:
I have no hesitation in saying that the work could not have been carried on continuously in these tropical seas for such a length of time in any other kind of ship. The principal points of advantage were, the thick wooden walls, which completely prevented over heating and over cooling, the splendid ventilation which was provided by the twenty-gun ports on the main deck, and the practice of making the passage under sail (42, pp.36–37).

189 Only in one case – the determination of potash – was there any deviation (79, p.3).

190 Italics are my emphasis.

191 Italics are my emphasis.

192 The accepted values for atomic weights had changed in the 20 years between the respective work of Forchhammer and that of Dittmar. In his treatise Dittmar gave the atomic weights he had used.

193 By total salts Dittmar meant the sum total of the major constituents rather than that as determined by evaporation of 1 kilogram sea water.

194 Notice the use of "chlorine" not "chlorinity." From Knudsen's tables a Chlorinity of 19.149% would be equivalent to a salinity of 34.60.

195 He, like Dittmar, when referring to chlorine meant the bromine as well.

196 These last two were taken to be synonymous, a result of Forchhammer's work.

197 This method was still not precise. If Tørnoe had not been such a competent chemist and exhibited such good laboratory technique, the results would have been much more varied.

198 My emphasis.

199 In 1883 Arrhenius presented his dissertation for the doctorate at the University of Upsala (305). A condensed version was published in 1887 (6).

200 The member countries now number 16 (73, p.47).

201 The names of those chosen are readily recognized by one familiar with the modern study of the sea. They were: Murray, Knudsen, Nansen, Krummel, Dickson and Makaroff.

202 He held this post until his death in 1949.

203 This was named after the Danish vessel "H.M.S. Ingolf." It was then still common to name expeditions after the ship used.

[204] My calculation from rest of Knudsen's data.

[205] The addition of an excess of silver nitrate and backtitrating this excess with a very dilute ammonium thiocyanate solution with ferric ammonium sulfate as an indicator. This procedure was Dittmar's (164, p.82).

[206] Not the famous laboratory of that well known Danish beer, Carlsberg, but a lab, which that company which has always been more than generous to the sciences, had endowed.

[207] "Chlormenge" – literally chlorine content or quantity – is synonymous in English with the word chlorinity. From this time on, there was the tendency in English to speak of chlorinity. This was not new. Forchhammer had put forth the same definition in 1865 (100). But now it was defined officially.

[208] Here is the most important reason for Sorensen's decision to gravimetrically determine chlorine.

[209] The phrase "dissolved in 1000 grammes of water" is a mistranslation. The German version read "in 1000 g Seewasser gelöst". The difference is important; only the latter has meaning for oceanographers (60, p.77).

[210] As a matter of interest, it might be added that the commission had instructed Knudsen to investigate other possible methods of measuring salinity. Knudsen did try electrical methods (conductivity) to determine this parameter but found them to be unreliable. This was in 1901 (166).

[211] Italics are mine.

[212] At least until 1966.

[213] Inconsistencies are not confined to older determinations. As an example of more recent inconsistencies, the phosphate determinations in the Norpac data, 1966, need only be pointed out (44).

[214] A variety of physical properties of sea water depend upon salinity. For example, a recent text states the following relationships: A. Increase with increasing salinity: density, molecular viscosity, surface tension, refractive index, electrical conductivity, coefficient of thermal expansion, speed of sound, osmotic pressure. B. Decrease with increasing salinity: specific heat, freezing point temperature, temperature of maximum density, vapor pressure, thermal conductivity (molecular) (192, p.20).

[215] S‰ = 1.805 Cl‰ + 0.030.

[216] This result may have been determined graphically. Mathematical calculation yields 35, not 34.993.

[217] These means are not under discussion here, but for those who wish to pursue them further, Chapter 3 of *Chemical Oceanography* (60), *The Physical Properties of Sea Water*, by R.A. Cox is strongly recommended.

[218] It consisted of: D.E. Carritt (IAPO), R.A. Cox (ICES), G. Dietrich (SCOR), N.P. Fofonoff (IAPO), F. Hermann (ICES), G.N. Ivanoff-Frantskevich (UNESCO), and Y. Miyake (SCOR) (292, pp.2–3).

[219] The remaining nine recommendations are given in Appendix IX.

[220] My emphasis.

[221] The acidification was with nitric acid.

[222] This involved the use of iron (II) sulfate. Today's method of Lajos Winkler uses manganese chloride (314).

[223] For example, in a work of 1863, the author stated in his partial analysis of sea water: "La richesse de cette eau en chlorure de sodium a été déterminée par la procédé de M. Mohr. La moyenne de deux analyses tout a fait concordantes m'a donne pour 1 litre de l'eau n° 1,33 gr, 950 de chlorure de sodium" (119, p.188). The French oceanographer Thoulet, in his book *L'Océan* (289) in 1904, had this to say about Mohr. "Le savant qui a le mieux compris le rôle de la chimie dans l'histoire

de la terre en géneral et dans celle de la mer en particulier, est Mohr, le chimiste éminent, le véritable propagateur pour ne pas dire le créateur des méthodes d'analyse chimique par les liqueurs titrées" (289, pp.71–72). He added further: "Je ne puis resister au plaisir de copier quelques-unes de ces thèses qui concernent spécialement la chimie de la mer. Je les laisse en latin et je bornerai mes citations afin de me faire pardonner mon admiration pour Mohr."

Thoulet then quoted the following passage in Latin which is here translated:

23. Mare est atmosphaera quaedam omnia continens ad vitam plantarum et animalium necessaria. (The sea is an atmosphere containing all thing necessary to the life of plants and animals.)

26. Quidquid calcariam carbonicam in mari producit est animal; montium calcareorum formatio sempiterna est et adhuc viget. (Whatever produces calcium carbonate in the sea is animal; the formation of calcareous mountains is continual and even now goes on.)

35. Acidi phosphorici circulation est ex aqua marina per animalium incrustationes in saxa calcarea, et inde in terram, montes, solum agreste et per fluvios reditus in mare. (The circulation of phosphoric acid is from sea water through incrustations of animals in calcareous rocks, and thence into earth, mountains, valleys and returns by rivers to the sea).

36. Fluorium, satelles acidi phosphorici, ex aqua marina per montes calcareos in terram dissipatur. (Fluorine, associated with phosphoric acid, from marine water is scattered by mountains in the earth.)

38. Magnesia omnis ex chlorido magnesico et sulphate magnesico aquae marinae derivanda. (All magnesia from magnesium chloride and magnesium sulphate must be derived from marine water.)

39. Sulfur et sulfuris combinationes in terra ad gypsum marinum referuntur. (Sulfur and combinations of sulfur in the earth are returned to marine gypsum.)

83. Non est neque minerale neque saxum quod dissolutioni in infinitum resistat. (It is not the case that mineral or rock resists dissolution to infinity.)

85. Hodierni temporis terra geologiam docet proeteriti. (The earth of the present time teaches the geology of the past.)

95. Chemica in rebus chemicis explicandis dux et magistra, et neglecta in geologis celeberrimis graviter vindicata est (289, pp.72–73). (Chemistry, in things chemical to be explained, is the leader and master, and the neglecting by celebrated geologists is to be vehemently vindicated.) These little known comments by Mohr on the sea appeared in *Geschichte der Erde*, first published in 1866 (215).

224 The word titrimetric was coined by Dittmar.

225 At almost the same time as the "Challenger" Expedition, the Norwegian North Atlantic Expedition (1876–1878), which actually was comprised of two separate expeditions, investigated the physical, zoological, and chemical aspects of the ocean between Norway, the Faroe Islands, Iceland, and Spitzbergen (270, p.2). The scientific results as it was a much smaller expedition were published in 1880–1882, several years before those of the "Challenger". These results were available to Dittmar prior to his writing of the chemical results for the *Challenger Reports* (79).

226 Roux said he used the Mohr method. He was the first analyst of sea water to cite the Mohr procedure.

227 F. Krüger, in 1876 (171), first used fluorescein as a titrimetric indicator although not in the chlorine determination (283, p.260).

BIBLIOGRAPHY

1. Agricola, Georgius. *De Re metallica* (translated by Herbert C. Hoover and Lou H. Hoover). Dover, New York, 1950, 617 pp.
2. Aimé, G., 1845. *Recherches de Physique sur la Méditerranée*. Imprimerie Royale, Paris, 211 pp.
3. Aimé, G., 1845. Mémoire sur les températures de la Méditerranée. *Ann. Chim. Phys., Ser. 3*, 15 : 5–34.
4. Allen, E.J., 1926. A selected bibliography of marine bionics and fishery investigation. *J. Conseil* 1–2 : 77–96.
5. Aristotle, Ross, W.D. (Editor), 1938. *The Works of Aristotle*, Clarendon, Oxford, 12 vols.
6. Arrhenius, S., 1887. Dissociation of substances dissolved in water. *Z. Phys. Chem.*, 1 : 631–648.
7. Arrhenius, S., 1889. Electrolytic dissociation versus hydration. *Phil. Mag. J. Sci., Ser.5*, 28 : 30–38.
8. Arrhenius, S., 1913. *Theories of Solutions.* Yale University, New Haven, 225 pp.
9. Bailey, W., 1582. *A Briefe Discours of Certain Bathes or Medicinall Waters in the Countie of Warwicke neere unto a Village Called Newnam Regis.* London [?], 35 pp.
10. Balard, Antoine-Jérome, 1822. Note pour servir a l'histoire naturelle de l'iode. *Ann. Chim. Phys., Ser.2*, 28 : 170–181.
11. Balard, Antoine-Jerome, 1826. Sur une substance particuliere contenue dans l'eau de la mer. *Ann. Chim. Phys., Ser.2*, 32 : 337–381.
12. Barth, T.F.W., 1952. *Theoretical Petrology*. Wiley, New York, N.Y., 416 pp.
13. Bein, W., Hirsekorn, H. and Möller, L., 1935. Konstantenbestimmungen des Meerwassers und Ergebnisse über Wasserkörper. *Berlin Univ. Inst. Meereskunde, Veröffentl., Neue Folge Geogr. Naturwiss. Reihe*, Heft 28 : 240 pp.
14. Bergman, Torbern, 1784. *Physical and Chemical Essays* (translated by Edmund Cullen) Murray, London, 2 vol.
15. Bergman, Torbern, 1956. *On Acid of Air, Treatise on Bitter, Seltzer, Spa and Pyrmont Waters and their Synthetical Preparation* (Edited by Uno Boklund and translated by Sven M. Jonsson. Almquist and Wiksell, Stockholm, 128 pp.
16. Bibra, Ernst von, 1851. Untersuchung von Seewasser des Stillen Meeres und des Atlantischen Oceans. *Justus Liebig's Ann. Chem.*, 77 : 90–102.
17. Bischof, Gustav., 1847–1855. *Lehruch der chemischen und physikalischen Geologie*. Marcus, Bonn, 2 vol.
18. Blagden, Sir Charles., 1788. Experiments on the effect of various substances in lowering the point of congelation in water. *Phil. Trans. Roy. Soc. Lond.*, 78 : 277–312.
19. Blagden, Sir Charles, 1788. Experiments on the cooling of water below its freezing point. *Phil. Trans. Roy. Soc. Lond.*, 78 : 129–130.
20. Black, Joseph, 1794. An analysis of the waters of some hot springs in Iceland. *Trans. Roy. Soc. Edinb.*, 3 : 95–121.
21. Black, Joseph, 1803. *Lectures on the Elements of Chemistry, Delivered in the University of Edinburgh by the Late Joseph Black, M.D., now Published from his Manuscripts.* Creech, Edinburgh, 2 vol.
22. Boas, Marie, 1956. Acids and alcalis in 17th century chemistry. *Arch. Int. Hist. Sci.*, 9 : 13–28.
23. Boas, Marie, 1958. *Robert Boyle and Seventeenth Century Chemistry*. Cambridge University, Cambridge, 239 pp.
24. Bouillon-Lagrange, E.J.B., 1811. Essai sur les eaux minérales naturelles et artificielles. *Bull. Pharmacie, Sér.1*, 3 : 47–48.

25. Bouillon-Lagrange, E.J.B. and Vogel, A., 1813. Mémoire sur l'eau des mers qui baignent les côtes de l'Empire français. *Bull. Pharmacie,* 5 : 505–516.
26. Boulduc, Gilles Francois, 1726. L'esprit-de-vin dans l'analyse. *Mém. Acad. Roy. Sci. (Paris),* 1726: 412.
27. Bouquet de la Grye, A., 1882. Sur la densité et la chloruration de l'eau de mer puissée à bord du "Travailleur" en 1881. *Compt. Rend. Acad. Sci. (Paris),* 94 : 1063–1066.
28. Bouquet de la Grye, A., 1882. Recherches sur la chloruration de l'eau de mer. *Ann. Chim. Phys.,* 25 : 433–477.
29. Boussingault, E.J.B., 1825. Sur l'existence de l'iode dans l'eau d'une saline de la province d'Antioquia. *Ann. Chim. Phys.,* 30 : 91–96.
30. Boussingault, E.J.B., 1832. Analyse de l'eau du Rio Vinagre. *Ann. Chim. Phys.,* 51 : 107–110.
31. Boussingault, E.J.B.,1856. Recherches sur les variations que l'eau de la Mer Morte paraît subir dans sa composition. *Ann. Chim. Phys., Sér.3,* 48 : 129–170.
32. Boussingault, E.J.B., 1857. Sur les quantités de nitrates contenues dans le sol et dans les eaux. *Compt. Rend. Acad. Sci. (Paris),* 44 : 108–119.
33. Boyle, Robert, 1693. An account of the honourable Robert Boyle's way of examining waters as to freshness and saltness. *Phil. Trans. Roy. Soc. Lond.,* 17 : 627–641.
34. Boyle, Robert, 1666. An account of a book, very recently published, entitled "The Origins of Forms and Qualities", illustrated with considerations and experiments. *Phil. Trans, Roy. Soc. Lond.,* 1 : 191–198.
35. Birch, T. (Editor), 1965. *The Works of Robert Boyle.* Georg Olms, Hildesheim, 6 vol.
36. Brock, W.H., 1967. An attempt to establish the first principles of the history of chemistry. *Hist. Sci.,* 6 : 156–169.
37. Browne, C.A., 1943. Thomas Jefferson's relation to chemistry. *J. Chem. Educ.,* 20 : 574.
38. Buchan, Alex, 1894–1896. Specific gravities and oceanic circulation. *Trans. Roy. Soc. Edinb.,* 38: 317–341.
39. Buchanan, J.Y., 1874–1875. On the determination, at sea, of the specific gravity of sea-water. *Proc. Roy. Soc. Lond.,* 23 : 301–308.
40. Buchanan, J.Y., 1884. *Report of the Specific Gravity Samples of Ocean Water, Observed on Board HMS "Challenger".* Edinburgh, 46 pp. (Challenger Office. Report on the scientific results of the voyage of HMS "Challenger" during the years 1873–1876. *Phys. Chem.,* 1.)
41. Buchanan, J.Y., 1912–1914. On the specific gravity and displacement of some saline solutions. *Trans. Roy. Soc. Edinb.,* 49 : 1–228.
42. Buchanan, J.Y., 1919. *Accounts Rendered of Work Done and Things Seen.* Cambridge University, Cambridge, 435 pp.
43. Calamai, 1848. Analyse des Meerwassers. *Z. Prak. Chem.,* 5 : 235–238.
44. California University, Scripps Institution of Oceanography, 1960. *Oceanic Observations of the Pacific: 1955. (The NORPAC data prepared by the NORPAC Committee)* Berkeley, 532 pp.
45. Campbell, W.A. and Mallen, C.E., 1958. The development of quantitative analysis from 1750 to 1850, 1. *Proc. Univ. Durham Phil. Soc.,* 13A : 108–118.
46. Campbell, W.A. and Mallen, C.E., 1960. The development of qualitative analysis (1750–1850), 2. *Proc. Univ. Durham Phil. Soc.,* 13A : 168–173. 1960.
47. Carpenter, W.B., 1872. Report on scientific researches carried on during the months of August, September, and October 1871 in H.M. surveying ship "Shearwater". *Proc. Roy. Soc. Lond.,* 1872 : 535–644.
48. Carritt, D.E., 1952. Chemical measurements. In: *Oceanographic Instrumentation – Natl. Acad. Sci. Natl. Res. Council, Publ.* 309 : 166–193.
49. Carritt, D.E., 1963. Chemical instrumentation. In: M.N. Hill (Editors), *The Sea, 2.* Interscience, London, 109–123.
50. Carrit, D.E. and Carpenter, J.H., 1958. The composition of sea water and the salinity–chlorinity–density problems. In: *Conference on Physical and Chemical Properties of Sea Water, Easton, Maryland, 1958 – Natl. Acad. Sci. Natl. Res. Council, Publ.,* 600 : 67–85.
51. Carruthers, J.N., 1963. Some oceanography from the past. *Inst. Nav. J.,* 16 : 180–188.
52. Cavendish, Henry, 1767. Experiments on Rathbone-Place water. *Phil. Trans. Roy. Soc. Lond.,* 57 : 92–108.

53. Charpentier, Paul, 1873. Nouvelles méthodes de dosage volumétrique du fer et des alcalis. *Rev. Univers. Mines. Metall. Trav. Publ. Sci. Arts, Appl. Ind.*, 32 : 302–306.
54. Chu, S.P., 1949. Experimental studies on the environmental factors influencing the growth of phytoplankton. *Sci. Technol. China*, 2 : 38–44.
55. Clemm, G., 1841. Analyse des Nordsee-Wassers. *Justus Liebig's Ann. Chem.*, 37 : 111–113.
56. Committee on chemical properties. Report, summary. In: *Conference on the Physical and Chemical Properties of Sea Water, Easton, Maryland, 1958 – Natl. Acad. Sci., Natl. Res. Counc., Publ.*, 600 : VII–IX.
57. Courtois, Bernard, 1813. Découverte d'un substance novelle dans le vareck. *Ann. Chim. Phys., Sér. 1*, 88 : 304–310.
58. Cowen, R.C., 1960. *Frontiers of the Sea.* Doubleday, New York, 299 pp.
59. Cox, R.A., 1963. The salinity problem. In: M. Sears (Editor), *Progress in Oceanography*, 1. MacMillan, New York, pp.243–261.
60. Cox, R.A., 1965. The physical properties of sea water. In: J.P. Riley and G. Skirrow (Editors), *Chemical Oceanography*, 1. Academic Press, New York, pp.73–120.
61. Cox, R.A., Culkin, F. and Riley, J.P., 1967. The electrical conductivity/chlorinity relationship in natural sea water. *Deep Sea Res.*, 14 : 203–220.
62. Cox, R.A., 1962. The chlorinity, conductivity and density of sea water. *Nature*, 193 : 518–520.
63. Culkin, F., 1965. The major constituents of sea water. In: J.P. Riley and G. Skirrow (Editors), *Chemical Oceanography*, 1. Academic Press, New York, pp.121–162.
64. Dalton, John, 1824. On the saline impregnation of the rain, which fell during the late storm, December 5th, 1822. *Mem. Manch. Lit. Phil. Soc., Ser. 2*, 4 : 324–329.
65. Darcet, Description. *Ann. Ind. Natl. Etrangère, Sér. 2*, 2 : 186–192.
66. Darondeau, 1838. Résultat de l'examen des eaux de mer recueillies pendent le voyage de la Bonite, avec l'appareil imaginé per M. Biot. *Ann. Chim. Phys. Sér. 2*, 69 : 100–106.
67. Darwin, Charles, 1962. *The Voyage of the Beagle.* Doubleday, Garden City, 507 pp.
68. Daubrée, M.A., 1851. Recherches sur la présence de l'arsénic et de l'antimoine dans les combustibles minéraux, dans diverses roches et dans l'eau de la mer. *Ann. Mines*, 19 : 669–683.
69. Davy, John, 1824. Observations on the specific gravity and temperature made during a voyage from Ceylon to England in 1819 and 1820. *Edinb. Phil. J.*, 10 : 317–322.
70. Davy, John, 1849. On carbonate of lime as an ingredientof sea-water. *Proc. Roy. Soc. Lond.*, 5 : 828–831.
71. Davy, John, 1861. On the rain-fall of the lake district. *Trans. Roy. Soc. Edinb.*, 23 : 53–65.
72. Day, F.H., 1963. *The Chemical Elements in Nature.* London, Harrap, London, 364 pp.
73. Dean, J.R., 1966. *Down to the Sea, a Century of Oceanography.* Brown, Son, and Ferguson, Glasgow, 122 pp.
74. Debus, A.G., 1962. Sir Thomas Browne and the study of colour indicators. *Ambix*, 10 : 29–36.
75. Debus, A.G., 1962. Solution analysis prior to Robert Boyle. *Chymia*, 8 : 41–61.
76. DeWitt, Benjamin, 1788. A memoir on the Onondaga salt springs and salt manufacturies in the western part of the State New York. In: *Additions and Corrections to the Former Additions of Dr. Robertson's History of America.* Cadell, London, pp.1–27.
77. Dietrich, G., 1962. Algemeine Meereskunde: Eine Einführung in die Ozeanographie. In: *Ozeanographische Instrumente und Messmethoden.* Borntraeger, Berlin, pp.80–124.
78. Dieulafait, Louis, 1877. L'acide borique. *Compt. Rend. Acad. Sci. (Paris)*, 85 : 605–607.
79. Dittmar, William, 1884. *Physics and Chemistry.* Edinburgh, 251 pp. (Challenger Office. Report on the scientific results of the voyage of HMS "Challenger" during the years 1873–76, 1).
80. Duclos, Samuel, 1733. Autres expériences de chimie. Analise de plusiers eaux. In: *Histoire de l'Académie des Sciences depuis 1666 jusqu'à son renouvellement en 1699.* Martin, Coignard and Guerin, Paris, pp.24–35.
81. Duclos, Samuel, 1733. Expérience pour dessaler l'eau de la mer. In: *Histoire de l'Académie des Sciences depuis 1666 jusqu'à son renouvellement en 1699.* Martin, Coignard and Guerin, Paris, pp.50–51.
82. Duclos, Samuel, 1733. Eaux minérales (1670). In: *Histoire de l'Académie des Sciences depuis 1666 jusqu'à son renouvellement en 1699.* Martin, Coignard and Guerin. Paris, pp.123–124.

83. Duclos, Samuel, 1733. Examen des eaux de Versailles (1683). In: *Histoire de l'Académie des Sciences depuis 1666 jusqu'à son renouvellement en 1699.* Martin, Coignard and Guerin, Paris, pp.367–369.

84. Duclos, Samuel, 1733. Sur une manière de dessaler l'eau de la mer (1684). In: *Histoire de l'Académie des Sciences depuis 1666 jusqu'à son revouvellement en 1699.* Martin, Coignard and Guerin, pp.387–389.

85. Duval, C., 1951. François Descroizilles, the inventor of volumetric analysis. *J. Chem. Educ.*, 28 : 508–519.

86. El-Mas'Udi, 1841. *Meadows of Gold and Mines of Gems.* (Translated by Aloys Sprenger.) Allen, London, 2 vol.

87. Fahrenheit, Daniel Gabriel, 1724. Araeometri novi descriptio et usus. *Phil. Trans. Roy. Soc. Lond.*, 33 : 140–141.

88. Ferguson, J., 1954. *Bibliotheca Chemica.* Derek Verschoyle, London, 2 vol.

89. Figuier, Louis and Miahle, L., 1848. Examen comparatif des principales eaux minérales salines d'Allemagne et de France, sous le rapport chimique et thérapeutique. *J. Pharmacie*, 13 : 401–411.

90. Fleming, John, 1818. Junction of the fresh waters of rivers with the salt water of the sea. *Trans. Roy. Soc. Edinb.*, 8 : 507–513.

91. Forch, C., 1902. Volumausdehnung des Seewassers. In: M. Knudsen (Editor), *Berichte uber die Konstantenbestimmungen zur Aufstellung der hydrographischen Tabellen, 4. – Kon. Danske Videnskab. Selsk. Skrifter, Naturvidenskab. Mathemat., Raekke 6*, 12 (1) : 139–152.

92. Forchhammer, Georg, 1845. Om sövandets sammensaetning, dets bestanddele: meddehavet og nordsven. *Kon. Danske Videnskab. Selsk., Oversigt,* 1845 : 24–33.

93. Forchhammer, Georg, 1846. On comparative analytical researches. *Rep. Brit. Assoc. Advan. Sci.*, 1846 : 90–91.

94. Forchhammer, Georg, 1846. Ueber die Bestandtheile des Meerwassers, seine Strömungen und deren Einfluss auf das Klima der Küsten von Nord-Europa. *Ges. Dtsch. Naturforsch. Aerzte*, 24 : 77–108.

95. Forchhammer, Georg, 1847. Composition de l'eau de la mer. *J. Pharmacie Chim., Sér.3,* 11 : 475–476.

96. Forchhammer, Georg, 1849–50. On a new method of determining the organic matter in water. *Chemist*, 1849–50 : 67–68.

97. Forchhammer, Georg, 1859. *Om Sövandets Bestanddele og deres Fordeling I havet.* Schultz, Copenhagen, 110 pp.

98. Forchhammer, Georg, 1861. Resultater af sine Undersogelser, saavel over Saltmaengden i middelhavets vand som over Forekomsten af Borsyre og leerjord: Sövandet. *Kon. Videnskab. Selsk., Oversigt,* 1861 : 379–391.

99. Forchhammer, Georg, 1862. On the constitution of sea-water, at different depths, and in different latitudes. *Proc. Roy. Soc. Lond.*, 12 : 129–132.

100. Forchhammer, Georg, 1865. On the composition of sea-water in the different parts of the ocean. *Phil. Trans. Roy. Soc. Lond.*, 155 : 203–262.

101. Fyfe, Andrew, 1819. On the quantity of saline matter in the water of the North Polar seas. *Edinb. Phil. J.*, 1 : 160–163.

102. Garrels, R.M. and Thompson, M.E., 1962. A chemical model for sea-water at 25°C and one atmosphere total pressure. *Am. J. Sci.*, 260 : 57–66.

103. Gay-Lussac, Joseph L., 1816. Description d'un thermomètre propre à indiquer des maxima ou des minima de température. *Ann. Chim.*, 3: 90–91.

104. Gay-Lussac, Joseph L., 1817. Note sur la salure de l'Océan atlantique. *Ann. Chim. Phys., Sér.2,* 6 : 426–436.

105. Gay-Lussac, Joseph L., 1817. Supplement à la note sur la salure de la mer, imprimée dans le sixième volume des Annales de Chimie et de Physique. *Ann. Chim. Phys., Sér.2,* 7 : 79–83.

106. Gay-Lussac, Joseph L., 1819. Premier mémoire sur la dissolubilité des sels dans l'eau. *Ann. Chim. Phys. Sér.2,* 11 : 296–315.

107. Gay-Lussac, Joseph L., 1832. *Instruction sur l'Essai des Matières d'Argent per la Voie humide.* Imprimérie Royale, Paris, 88 pp.

108. Gay-Lussac, Joseph L., 1835. Nouvelle instruction sur la chlorométrie. *Ann. Chim.*, 60 : 218–224.
109. Gay-Lussac, Joseph L., 1836. Nouvelle instruction sur la chlorométrie. *Ann. Chim.*, 63 : 334–335.
110. Gay-Lussac, J.L. and Humboldt, F.H. Alexander, 1805. Expériences sur les moyens eudiométriques et des principes constituants de l'atmosphère. *Ann. Chim.*, 53 : 239–259.
111. Gay-Lussac, Joseph L. and Thenard. L.J., 1808. Sur la décomposition et la récomposition de l'acide boracique. *Ann. Chim. Phys., Ser.1*, 68 : 169–174.
112. Gionetti, Victor Aimé, 1779. *Analyse des eaux minérales de S. Vincent et de Courmayeur dans le Duche d'Aoste avec une appendice sur les eaux de la Saxe, de Pré S. Didier et de Fontane-More.* Turin, 98 pp.
113. Gmelin, Christian Gottlob, 1827. Chemische Untersuchung des Wassers vom Todten Meere. *Ann. Chim.*, 35 : 102–105.
114. Göbel, Friedemann, 1842. Resultate der Zerlegung des Wassers vom Schwarzen, Asowschen und Kaspischen Meeres. *Ann. Phys. Chem.*, 51 : 187–188.
115. Göbel, Friedemann, 1838. *Reise in die Steppen des südlichen Russlands.* Kluge, Dorpat, 372 pp.
116. Goldberg, E.D., 1960. Chemists and the oceans. *Chymia*, 6 : 162–179.
117. Goldberg, E.D., 1965. Minor elements in sea water. In: J.P. Riley and G. Skirrow (Editors), *Chemical Oceanography,* 1. Academic Press, New York, pp.163–196.
118. Graham, R.P., 1947. Some definitions in quantitative analysis. *J. Chem. Educ.*, 24 : 596–599.
119. Grandeau, Louis, 1863. Sur la présence du rubidium et du caesium, dans les eaux naturelles, les minéraux et les végétaux. *Ann. Chim. Phys., Sér.3*, 67 : 155–236.
120. Groen, P., 1967. *Waters of the Sea.* Van Nostrand, London, 323 pp.
121. Guberlet, M.L., 1964. *Explorers of the Sea; Famous Oceanographic Expeditions.* Ronald, New York, 226 pp.
122. Guntz, A.A. and Kocher, J., 1952. Mesure de la salinité des eaux de mer. *Compt. Rend. Acad. Sci., (Paris) Sér.23*, 234 : 2300–2302.
123. Hall, A.R., 1956. *The Scientific Revolution.* Beacon, Boston, 379 pp.
124. Halley, Edmund, 1687. An estimate of the quantity of vapour raised out of the sea by the warmth of the sun; derived from an experiment shown before the Royal Society, at one of their late meetings. *Phil. Trans. Roy. Soc. Lond.*, 16 : 366–370.
125. Halley, Edmund, 1691. An account of the circulation of the watery vapours of the sea, and of the cause of springs, presented to the Royal Society. *Phil. Trans. Roy. Soc. Lond.*, 16 : 468–473.
126. Halley, Edmund, 1693. An account of the evaporation of water, as it was experimented in Gresham College in the year 1693. With some observations thereon. *Phil. Trans. Roy. Soc. Lond.*, 28 : 183–185.
127. Halley, Edmund, 1714. Some remarks on the variations of the magnetical compass published in the memoirs of the Royal Academy of Sciences, with regard to the general chart of those variations made by E. Halley; as also concerning the true longitude of the Magellan Streights. *Phil. Trans. Roy. Soc. Lond.*, 28 : 165–168.
128. Halley, Edmund, 1715. A short account of the cause of the saltness of the ocean, and of the several lakes that emit no rivers; with a proposal, by help thereof, to discover the age of the world. *Phil. Trans. Roy. Soc. Lond.*, 29 : 296–300.
129. Hampel, C.A., 1948. Fresh water from the sea. *Chem. Eng. News*, 26 : 1982–1985.
130. Harvey, H.W., 1963. *The Chemistry and Fertility of Sea Waters.* Cambridge University, Cambridge, 234 pp.
131. Hauksbee, Francis, 1709. An account of an experiment touching the different densities of common water, from the greatest degree of heat in our climate, to the freezing point, observ'd by a thermometer. *Phil. Trans. Roy. Soc. Lond.*, 26 : 267–270.
132. Hayes, August, A., 1851. On the different chemical conditions of the water at the surface of the ocean and at the bottom, on soundings. *Am. J. Sci., Ser.2,* 11 : 241–244.
133. Henry, Thomas, 1781–1783. On the natural history of magnesium earth. *Mem. Manch. Lit. Phil. Soc., Ser.1,* 1 : 442–473.
134. Henry, Thomas, 1781–1783. On the preservation of sea water from putrefaction by means of quicklime. *Mem. Manch. Lit. Phil. Soc., Ser.1,* 1 : 41–54.

135. Herdman, W.A., 1923. *Founders of Oceanography and Their Work.* Arnold, London, 340 pp.
136. Hodgman, C. (Editor), 1955. *Handbook of Chemistry and Physics.* Chemical Rubber, Cleveland, 37th ed., 3116 pp.
137. Hoek, P.P.C., 1903. Report of administration for the first year: 22d July 1902–21st July 1903, Charlottenlund Slot, Denmark, 1903. *Conseil Perm. Int. Explor. Mer, Rapp.* 1 : 39 pp.
138. Hoff, J.H. van 't., 1905. *Zur Bildung der Ozeanischen Salzablagerungen.* Vieweg, Braunschweig, 171 pp.
139. Home, Francis, 1756. *An Essay on the Contents and Virtues of Dunse Spaw.* Kincaid and Donaldson, Edinburgh, 330 pp.
140. Hooke, Robert, 1667. To fetch up water from any depth of the sea. *Phil. Trans. Roy. Soc. Lond.*, 2 : 447–448.
141. Hope, Thomas Charles, 1839. Inquiry whether sea-water has its maximum density a few degrees above its freezing point, as pure water has. *Trans. Roy. Soc. Edinb.*, 14 : 242–252.
142. Horner, Johann Kaspar, 1819. Ueber das specifische Gewicht des Meerwassers in verschiedenen Gewässern. *Ann. Phys.*, 63 : 159–189.
143. Horner, Johann Kaspar, 1820. Einige Bemerkungen zu den Aufsatzen über das Meerwasser vom jetzigen Winter. *Ann. Phys.*, 64 : 98–102.
144. Humboldt, Friedrich Wilhelm Heinrich Alexander von, 1804. Notes sur son voyage en Amérique. *Ann. Museum Natl. Hist. Nat.*, 3 : 396–404.
145. Humbolt, Friedrich Wilhelm Heinrich Alexander von. 1894–1900. *Personal Narrative of Travels to the Equinoctial Regions of America, During the Years 1799–1804.* Bell, London, 2 vol.
146. Hutton, James, 1792. *Dissertations on Different Subjects in the Natural Philosophy.* Strahan and Cadell, London, 686 pp.
147. Hutton, James, 1795. *Theory of the Earth.* Creech, Edinburgh, 2 vol.
148. Jacobson, J.P. and Palitzsch, S., 1921. Manuel pratique de l'analyse de l'eau de mer. *Bull. Inst. Oceanogr. Monaco, Pap.* 390 : 1–16.
149. Jacobsen, J.P. and Palitzsch, S., 1922. Manuel pratique de l'analyse de l'eau de mer. *Bull. Inst. Oceanogr. Monaco, Pap.* 409 : 1–31.
150. Jacobson, J.P. and Knudsen, M., 1940. Urnormal 1937 or primary standard sea water 1937. *Int. Union Geodesy Geophys. – Assoc. Phys. Oceanogr., Publ. Sci.*, 7 : 38 pp.
151. Jahns, P., 1961. *Scientists of the Civil War.* Hastings House, New York, 284 pp.
152. Jefferson, T., 1943. Report on the method of obtaining fresh water from salt. *J. Chem. Educ.*, 20: 575–576.
153. Jenkins, J.T., 1921. *A Textbook of Oceanography.* Constable, London, 202 pp.
154. Johnson, Cuthbert William, 1826. *Observations on the Employment of Salt in Agriculture and Horticulture with Directions for its Application, Founded on Practice.* Longman, London, 147 pp.
155. Johnstone, J., 1926. *A Study of the Oceans.* Arnold, London, 208 pp.
156. Jorden, Edward, 1631. *A Discourse of Naturall Bathes, and Minerall Waters.* Harpel, London, 92 pp.
157. Kirwan, Richard, 1799. *An Essay on the Analysis of Mineral Waters.* Myers, London, 194 pp.
158. Klaproth, Martin H., 1813. Chemische Untersuchung des Wassers vom Todten Meere. *Ann. Phil.*, 1 : 36–39.
159. Knudsen, M., 1902. Einleitung. In: *Berichte über die Konstantenbestimmungen zur Aufstellung der hydrographischen Tabellen,* 1. Copenhagen, pp.5–14.
160. Knudsen, M., 1902. Einsammeln und Aufbewahren der Wasserprobleme. In: *Berichte über die Konstantenbestimmungen zur Aufstellung der hydrographischen Tabellen, 1.* Copenhagen, pp.15–28.
161. Knudsen, M. 1902. Bestimmung des spezifischen Gewichtes. In: *Berichte über die Konstantenbestimmungen zur Aufstellung der hydrographischen Tabellen, 2.* Copenhagen, pp.29–92.
162. Knudsen, M., 1903. Gefrierpunkttabelle für Meerwasser. *Conseil Perm. Int. Explor. Mer, Publ. Circonst.*, 5 : 11–13.
163. Knudsen, M., 1901. Report. (2me Conf. Int. Explor. Mer.) *Kristiania,* 1901 : 38 pp.
164. Knudsen, M. 1899. *Hydrography, 2.* Luno, Copenhagen, 128 pp.

165. Knudsen, M., (Editor), 1931. *Hydrographical Tables According to the Measurings of Carl Forch, J.P. Jacobsen, Martin Knudsen and S.P.L. Sorensen,* Tutein and Koch, London, 2nd ed., 63 pp.
166. Knudsen, M., 1900. *Maaling of Hawandets Temperatur og Saltholdighed ved Hjoelp af Elektrisk Telefombro. Beretning fra Kommissionen for videnskabelig Undersogelse af de Danske Farvande.* Aamodts, Kjobenhavn, 10 pp.
167. Knudsen, M., Forch, C. and Sorensen, S.P.L., 1902. Bericht über die chemische und physikalische Untersuchung des Seewassers und die Aufstellung der neuen hydrographischen Tabellen. In: *Wissenschaftliche Meeresuntersuchungen.* (Neue Folge, Band 6.) Lipsius and Tischer, Kiel, pp. 123–184.
168. Koettstorfer, J., 1878. Zum Nachweis von Jod im Meerwasser. *Z. Anal. Chem.*, 17 : 305–309.
169. Koninck, L.L. de, 1901. Historique de la méthode titrimétrique. *Bull. Soc. Chem. Belg.*, 15 : mm173. 28–40; 73–90.
170. Kopp, Herman, 1844. *Geschichte der Chemie.* Vieweg, Braunschweig, 2 vol.
171. Krüger, Friedrich, 1876. Fluorescein als Indicator beim Titriren. *Ber. Dtsch. Chem. Gest.*, 9 : 1572.
172. Kuhn, T.S., 1964. *The Structure of Scientific Revolutions.* Phoenix, Chicago, 172 pp.
173. Laurens, Auguste, 1835. Sur une analyse de l'eau de la mer Mediterranée. *J. Pharmacie, Sér.2,* 21 : 89–94.
174. Lavoisier, Antoine, 1772. Memoire sur l'usage de l'esprit-de-vin dans l'analyse des eaux minerales. *Mém. Acad. Roy. Sci. (Paris)* 1772 : 555–563.
175. Lavoisier, Antoine. *Oeuvres de Lavoisier.* Johnson, New York, 1965, 6 vol.
176. Leicester, H.M., 1956. *The Historical Background of Chemistry.* Wiley, New York, 241 pp.
177. Lenz, Emil, 1830. Ueber das Wasser des Weltmeers in verschiedenen Tiefen, in Rücksicht auf die Temperatur und den Salzgehalt. *Ann. Phys. Chem.*, 20 : 73–130.
178. Lenz, Emil, 1832. On the comparative quantity of salt contained in the waters of the ocean. *Edinb. J. Sci.*, 6 : 341–234.
179. Lewis, C.L., 1927. *Matthew Fontaine Maury – the Pathfinder of the Seas.* U.S. Naval Institute, Annapolis, 252 pp.
180. Lichtenberg, F.D. von, 1811. Chemische Untersuchung des Ostsee-Wassers. *J. Chem. Phys.*, 2 : 252–257.
181. Liebmann, A.J., 1956. History of distillation. *J. Chem. Educ.*, 35 : 166–173.
182. Lister, Dr., 1684. Some experiments about freezing, and the difference betwixt common fresh water ice, and that of sea water: also a probable conjecture about the original of the nitre of Aegypt. *Phil. Trans. Roy. Soc. Lond.*, 4 : 836–838.
183. Lüning, O., 1911. Ueber die Entstehung der Massanalyse. *Apothekov Z.*, 26 : 702–713.
184. Lyman, J., 1954. History of oceanography. In: E. Long (Editor), *Ocean Sciences*, U.S. Naval Institute, Annapolis, pp 12–25.
185. Lyman, J., 1958. Chemical considerations. In: *Conference on Physical and Chemical Properties of Sea Water, Easton, Maryland, 1958 – Natl. Acad. Sci. Natl. Res. Council, Publ.*, 600 : 87–97.
186. Lyman, J. and Fleming, R.H., 1940. Composition of sea water. *J. Marine Res.*, 3 : 134–146.
187. Lyman, J., Barquist, R.F. and Wolf, A.V., 1958. A new method for direct salinity determination. *J. Marine Res.*, 17 : 335–340.
188. Macadam, Stevenson, 1852. On the presence of iodine in various plants, with some remarks on its general distribution. *Chem. Gaz.*, 10 : 281–286.
189. MacCurdy, E. (Editor and Translator), 1939. *The Notebooks of Leonardo Da Vinci.* Reynal and Hitchcock, New York. 2 vol.
190. McKie, D., 1935. *Antoine Lavoisier, the Father of Modern Chemistry.* Lippincott, New York, 303 pp.
191. McKie, D., 1952. *Antoine Lavoisier: Scientist, Economist, Social Reformer.* Schuman, New York, 440 pp.
192. McLellan, H.J., 1965. *Elements of Physical Oceanography.* Pergamon, London, 146 pp.
193. Macquer, P.J., 1762–1765. Miscellanea taurinensia. *Mélanges Philos. Mathémat. Soc. Roy. Turin,* 3 : 1–17.
194. Macquer, P.J., Lavoisier A. and Sage, A., 1778. Analyse de l'eau du Lac Asphalite. *Mem. Acad. Roy. Sci. (Paris)*, 1778 : 69–72.

195. Madsen, E.R., 1958. *The Development of Titrimetric Analysis till 1806.* Gad, Copenhagen, 239 pp.
196. Malaguti, F.J., Durocher, J. and Sarzeaud, 1850. Sur la présence du plomb, du cuivre et de l'argent dans l'eau de la mer, et sur l'existence de ce dernier métal dans les plantes et les êtres organisés. *Ann. Chim. Phys., Sér.3,* 28 : 129–163.
197. Malaguti, F.J. and Durocher, J., 1859. Observations relatives à la présence de l'argent dans l'eau de la mer. *Compt. Rend. Acad. Sci. (Paris),* 49 : 536.
198. Mangelsdorf, P.C., Jr., 1967. Salinity measurements in estuaries. In: G.H. Lauff (Editor), *Estuaries – (Am. Assoc. Advan. Sci., Publ.,* 83) : 71–79.
199. Marcet, Alexander, 1807. An analysis of the waters of the Dead Sea and the River Jordan. *Phil. Trans. Roy. Soc. Lond.,* 97 : 296–314.
200. Marcet, Alexander, 1813. Observations on Mr. Klaproth's analysis of the water of the Dead Sea. *Ann. Phil.,* 1 : 132–135.
201. Marcet, Alexander, 1819. On the specific gravity, and temperature, in different parts of the ocean, and in particular seas; with some account of their saline contents. *Phil. Trans. Roy. Soc. Lond.,* 109 : 161–208.
202. Marcet, Alexander, 1820. Sur la pesanteur spécifique de l'eau de l'océan, dans diverses régions. *J. Pharmacie, Sér.2,* 6 : 380–381.
203. Marcet, Alexander, 1822. Some experiments and researches on the saline contents of sea-water, undertaken with a view to correct and improve its chemical analysis. *Phil. Trans. Roy. Lond.,* 112 : 448–456.
204. Marchand, Eugène, 1852. Sur la constitution physique et chimique des eaux naturelles. *Compt. Rend. Acad. Sci. (Paris),* 34 : 54–56.
205. Marchand, Eugène, 1855. Des eaux potables en général considerées dans leur constitution physique et chimique et dans leurs rapports avec la physique du globe, la géologie, la physiologie générale et l'hygiène publique, ainsi que dans leur applications à l'industrie et à l'agriculture; en particulier des eaux utilisées dans les arrondissements du Harvre et d'Yvetot. *Mém. Acad. Médecine (Paris),* 19 : 121–318.
206. Marchand, Eugène, 1865. Composition de l'eau de la mer, puisse dans la Manche. *J. Pharmacie,* 1 : 381.
207. Mariotte, Edmé, 1686. Traité du mouvement des eaux et des autres corps fluides, divisé en V parties. E. Michallet, Paris, 408 pp.
208. Mariotte, Edmé, 1717. *Oeuvres.* Leide, P. Vander, 2 vol.
209. Marsilli, Luigi F., 1711. *Brieve Ristretto del Saggio Fiscio Intorno alla Storia del Mare.* Andrea Poletti, Venice, 78 pp.
210. Marsilli, Luigi F., 1725. *Histoire Physique de la Mer.* De'pens, Amsterdam, 173 pp.
211. Maury, Matthew F., 1855. *The Physical Geography of the Sea.* Harper, New York, 360 pp.
212. Maury, Matthew F., 1861. *The Physical Geography of the Sea.* Sampson Low, London, 6th ed., 370 pp.
213. Mill, Hugh Robert, 1891. The Clyde Sea area. *Trans. Roy. Soc. Edinb.,* 36 : 641–729.
214. Mohr, Carl Friedrich, 1856. Neue massanalytische Bestimmung des Chlors in Verbindungen. *Ann. Chem. Pharmacie,* 97 : 335–338.
215. Mohr, Carl Friedrich, 1866. *Geschichte der Erde.* Cohen, Bonn, 554 pp.
216. Mohr, Carl Friedrich, 1877. *Lehrbuch der chemisch-analytischen Titrimethode.* Vieweg, Braunschweig, 760 pp.
217. LeMonnier, Louis Guillaume, 1747. L'analyse des eaux minérales. *Mém. Acad. Roy. Sci. (Paris),* 1747 : 259–271.
218. Moray, R., 1667. Observations made by a curious and learned person, sailing from England, to the Caribe-Islands. *Phil. Trans. Roy. Soc. Lond.,* 2 : 494–509.
219. Morren, Auguste, 1884. Sur les gaz que l'eau de mer peut tenir en dissolution en différents moments de la journée, et dans les saisons diverses de l'année. *Ann. Chim. Phys., Ser.3,* 12 : 5–56.
220. Morris, A.W. and Riley, J.P., 1964. The direct gravimetric determination of the salinity of sea water. *Deep Sea Res,* 11 : 899–904.

221. Muir, M.M.P., 1909. *History of Chemical Theories and Laws.* Wiley, New York, 567 pp.
222. Multhauf, R., 1954. John of Rupescissa and the origin of medical chemistry. *Isis*, 45 : 359–367.
223. Multhauf, R., 1956. The significance of distillation in Renaissance medical chemistry. *Bull. Hist. Medicine,* 30 : 329–346.
224. Murchison, Roderick I., 1865. Obituary - Vice-Admiral Robert FitzRoy. *J. Roy. Geogr. Soc. Lond.*, 35 : cxxviii–cxxxi.
225. Murchison, Roderick I., 1866. Obituary - Forchhammer. *J. Roy. Geogr. Soc. Lond.*, 36 : cxxxvi–cxxxviii.
226. Murray, John, 1818. An analysis of sea-water; with observations on the analysis of salt-brines. *Trans. Roy. Soc. Edinb.*, 8 : 205–244.
227. Murray, John, 1818. A general formula for the analysis of mineral waters. *Trans. Roy. Soc. Edinb.*, 8 : 359–279.
228. Murray, Sir John, 1884. Summary of the scientific results. In: *Report of the scientific results of the voyage of HMS "Challenger" during the years 1873–76,1.*
229. Murray, Sir John, 1893. On the chemical changes which take place in the composition of the sea-water associated with blue muds on the floor of the ocean. *Trans. Roy. Soc. Edinb.*, 37 : 481–507.
230. Murray, Sir John, 1910. *The Oceans.* Holt, New York, 248 pp.
231. Nairne, Edward, 1776. Experiments on water obtained from the melted ice of sea-water, to ascertain whether it be fresh or not; and to determine its specific gravity with respect to other water. Also experiments to find the degree of cold in which sea-water begins to freeze. *Phil. Trans. Roy. Soc. Lond.*, 66 : 249–256.
232. Natterer, Konrad, 1892. Chemische Untersuchungen im östlichen Mittelmeer. *Monatsh. Chem.*, 13 : 873–897.
233. Natterer, Konrad, 1893. Chemische Untersuchungen im östlichen Mittelmeer. *Monatsh. Chem.*, 14 : 624–673.
234. Natterer, Konrad, 1894. Chemische Untersuchungen im östlichen Mittelmeer. *Monatsh. Chem.*, 15 : 530–604.
235. *Ocean and oceanography*, 1966. In: *The Encyclopaedia Britannica*, 16, pp. 837–851.
236. Oldenburg, Henry, 1666. Some considerations touching a letter in the Journal des Scavans of May 24, 1666. *Phil. Trans. Roy. Lond.*, 1 : 228–230.
237. Oxner, M., 1920. Manuel pratique de l'analyse de l'eau de mer, *Bull. Comm. Int. Explor. Sci. Mer Médit.*, 3 : 1–28.
238. Paris, Société Royale de Médicine, 1779. *Histoire Année 1776.* École de Santé, Paris, 431 pp.
239. Partington, J.R., 1960. *A Short History of Chemistry.* Harper, New York, 385 pp.
240. Partington, J.R., 1961. *A History of Chemistry.* MacMillan, London, 5 vol.
241. Peligot, Eugène, 1855. Études sur la composition des eaux. *Ann. Chim., Sér.3,* 44 : 257–313.
242. Perrault, Pierre, 1680. A particular account, given by an anonymous French author in his book of the origins of fountains. printed at Paris: to shew, that the rain and snow-waters are sufficient to make fountains and rivers run perpetually. *Phil. Trans. Roy. Soc. Lond.*, 10 : 447–450.
243. Pettersson, Otto, 1894. A review of Swedish hydrographic research in the Baltic and North Seas. *Scot. Geogr. Mag.*, 10 : 281–302; 352–359; 413–427; 449–462; 525–539; 617–653.
244. Pfaff, Christian H., 1814. Ueber die Mischung des Ostseewassers in Hafen von Kiel, und die Coexistenz von Salzen, die sich wechselseitig zersetzen in Mineralwasser. *J. Chem. Phys.*, 11 : 8–15.
245. Pfaff, Christian H., 1818. Nachtrag zu meiner Analyse des Ostseewassers. *J. Chem. Phys.*, 22 : 271–273.
246. Pfaff, Christian H., 1821. *Handbuch der analytischen Chemie für Chemiker, Staatsärzte, Apotheker, Oekenomen und Bergwerkskündige.* Hammrich, Altoona, 2 vol.
247. Pfaff, Christian H., 1822. Ueber die Wollaston-Murray'sche Methode, die Talkerde aus salzigen Auflösungen zu scheiden. *J. Chem. Phys.*, 35 : 428–430.
248. Pfaff, Christian H., 1823. Ueber die Verstarkung des Salzgehalts des Meerwassers in der Tiefe durch das Gefrieren. *Ann. Phys.*, 75 : 363–367.
249. Pfaff, Christian H., 1825. Ueber die Kalisälze, den Salmiak- und Iod-Gehalt des Ostseewassers, so wie über die Doppelsälze aus Talkerde und Kali. *Jahrb. Chem. Phys.*, 15 : 378–382.

250. Pilkington, R., 1959. *Robert Boyle.* John Murray, London, 171 pp.

251. Pisani, F., 1855. Analyse de l'eau du Bosphore, pris à Bujuk-Dere, près l'embouchure de la Mer Noire. *Compt. Rend. Acad. Sci (Paris)*, 41 : 532.

252. Plakhotnik, A.F., 1961. History of the study of oceans and seas. *Tr. Inst. Istorii Estestvoznanivai Tekn.*, 37(2) : 57–79.

253. Pliny, *Natural History* (translation 1944 by H. Rackham), Harvard University, Cambridge, 10 vol.

254. Plot, Robert, 1685. De origine fontium tentamen philosophicum in praelectione habita coram Societate Philosophica nuper Oxoniae instituta ad scientiam naturalem promovendam. *Phil. Trans. Roy. Soc. Lond.*, 15 : 862–865.

255. Porta, J.B., 1957. *Natural Magick.* Basic Books, New York, 409 pp.

256. Quinton, René, 1904. *L'eau de Mer, Milieu organique.* Masson, Paris, 466 pp.

257. Read, J., 1963. *Through Alchemy to Chemistry.* Harper and Row, New York, 196 pp.

258. Riley, J.P., 1965. Historical introduction. In: J.P. Riley and G. Skirrow (Editors), *Chemical Oceanography, 1.* Academic Press, New York, pp.1–42.

259. Rose, Heinrich, 1835. Ueber die Zuzammensetzung des Wassers vom Elton-See im Asiatischen Russland, verglichen mit dem Meerwasser und dem Wasser vom Caspischen Meere. *Ann. Phys. Chem.*, 35 : 169–187.

260. Rose, Heinrich, 1851. Ueber den Einfluss des Wassers bei chemischen Zersetzungen. *Ann. Phys. Chem.*, 83 : 417–452.

261. Rouelle, G.F., 1777. Sur le sel marin. *J. Médecine*, 48 : 77.

262. Roux, Benjamin, 1863. Sur la composition de l'eau de la Mer Morte. *Compt. Rend. Acad. Sci. (Paris)*, 57 : 602–604.

263. Roux, Benjamin, 1864. Étude sur l'eau de l'océan. *Arch. Médicine Nav. Pharmacie Nav.*, 2 : 418–430.

264. Roux, Benjamin, 1864. Examen de l'eau de l'océan. *Ann. Chim. Phys., Sér.4*, 3 : 441–445.

265. Roux, Benjamin, 1864. Sur la salure de l'océan. *Compt. Rend. Acad. Sci. (Paris)*, 59 : 379–380.

266. Roux, Benjamin, 1875. Observations sur les sels marins. *Moniteur Sci.*, 17 : 657–714.

267. Rutty, J., 1757. *A Methodical Synopsis of Mineral Waters, Comprehending the Most Celebrated Medicinal Waters, Both Cold and Hot, of Great Britain, Ireland, France, Germany, and Italy, and Several Other Parts of the World.* Johnston, London, 668 pp.

268. Salinity in the ocean. In: *The Encyclopedia of Oceanography, 1.* New York, pp.758–760.

269. Sarton, G., 1957. *Six Wings: Men of Science in the Renaissance.* World, Cleveland, 309 pp.

270. Schmelk, Ludwig, 1882. *Chemistry. (Norwegian North Atlantic Expedition, 1876–1878, 1(9)* Grøndahl, Christiania, 71 pp.

271. Schweitzer, Gustavus, 1836. Analysis of sea-water as it exists in the English Channel near Brighton. *Phil. Mag.*, 8 : 267–270.

272. Scott, J.M., 1950. Karl Friedrich Mohr, 1806–1879. Father of volumetric analysis. *Chymia*, 3 : 191–201.

273. Singer, C., 1931. *The Story of Living Things.* Harper, New York, 568 pp.

274. Smeaton, W.A., 1954. The contribution of P.J. Macquer, T.O. Bergman and L.B. Guyton de Morveau to the reform of chemical nomenclature. *Ann. Sci.*, 10 : 87–106.

275. Smyth, William Henry, 1854. *The Mediterranean, a Memoir, Physical, Historical, and Nautical.* Parker, London, 520 pp.

276. Sonstadt, Edward, 1872. On the presence of iodate of calcium in sea water. *Chem. News*, 25 : 196–198; 231–232, 241–242.

277. Sorensen, S.P.L., 1902. Chlor- und Salzbestimmung. In: M. Knudsen (Editor), *Berichte über die Konstantenbestimmungen zur Aufstellung der hydrographischen Tabellen, 3.* Copenhagen, pp. 93–138.

278. Spratt, Thomas A.B., 1859. Sondes faites dans la Mer Méditerranée, entre Malte et Matapan en 1856–7, avec observations sur les meilleurs moyens à employer pour sonder à de grandes profoundeurs. *Ann. Hydrogr.*, 16 : 303–333.

279. Stillman, J.M., 1924. *The Story of Early Chemistry.* Appleton, New York, 438 pp.

280. Strickland, J.D.H. and Parsons, T.R. 1965. A manual of sea water analysis. *Can. Fish. Res. Board, Bull.*, 125 : 204 pp.

281. Sudhoff, K., 1922. *Kudzen Handbuch der Geschichte der Medicin.* Pagels, Berlin, 534 pp.
282. Sverdrup, H.U., Johnson, M.W. and Fleming, R.H., 1942. *The Oceans.* Prentice-Hall, Englewood Cliffs, N.Y., 1060 pp.
283. Szabadvary, F., 1966. *History of Analytical Chemistry,* translation by Gyula Svehla. Pergamon, London, 401 pp.
284. Thompson, T.G., 1958. A short history of oceanography with emphasis on the role played by chemistry. *J. Chem. Educ.,* 35 : 108–112.
285. Thompson, T.G. and Wright, C.C., 1930. Ionic ratios of the waters of the North Pacific Ocean. *J. Am. Chem. Soc.,* 52 : 915–921.
286. Thorndike, L., 1934. *A History of Magic and Experimental Science.* MacMillan, New York, 8 vol.
287. Thorpe, T.E. and Morton, E.H., 1871. On the composition of the water of the Irish Sea. *Mem. Manch. Lit. Phil. Soc., Ser.3,* 4 : 287–302.
288. Thorpe, Thomas Edward and Rucker, Arthur William, 1877. On the expansion of sea-water by heat. *Phil. Trans. Roy. Soc. Lond.,* 166 : 405–420.
289. Thoulet, J., 1904. *L'Océan, ses Lois et ses Problèmes.* Hachette, Paris, 382 pp.
290. Thoulet, J., 1922. *L'Océanographie.* Gauthier-Villars, Paris, 284 pp.
291. Tørnoe, Hercules, 1880. *Chemistry. (Norwegian North Atlantic Expedition, 1876–1878, 1).* Grøndahl, Christiania, 76 pp.
292. United Nations Educational Scientific and Cultural Organization, 1962. *Joint Panel on the Equation of State of Sea Water.* Paris, 14 pp.
293. United Nations Eductional Scientific and Cultural Organization. Joint Panel on Oceanographic Tables and Standards, 1966. *International Oceanographic Tables,* New York, 118 pp.
294. Usiglio, J., 1849. Analyse de l'eau de la Mediterranée sur les cotes de France. *Ann. Chim. Phys., Sér.3,* 27 : 92–107.
295. Usiglio, J., 1849. Etudes sur la composition de l'eau de la Mediterranée et sur l'exploitation des sels qu'elle contient. *Ann. Chim. Phys., Sér.3,* 27 : 172–191.
296. Vauquelin, Louis Nicolas, 1825. Sur le savon; et de l'action que quelques sels neutres exercent sur la solution de cette matière. *J. Pharmacie, Sér.2,* 11 : 497–505.
297. Vauquelin, Louis Nicolas and Thénard, Louis, 1813. D'un rapport fait à l'institut, le 6 mars 1813, sur l'analyse de eaux minérales de provins. *Bull. Pharmacie,* 5 : 369–376.
298. Venel, Gabriel François, 1755. L'analyse de l'eau minérale de Seltz. (Mémoires présentés par divers savans à l'Académie Royale des Sciences de l'Institute de France, 2 : 80–112.)
299. Vincent, 1862. Recherches sur l'eau de mer. *Ann. Chim. Phys., Sér.3,* 64 : 345–359.
300. Vogel, A., 1817. Analyse des Seewassers aus dem Canal, dem Atlantischen und Mittelländischen Meere. *J. Chem. Phys.,* 8 : 344–351.
301. Volhard, Jakob, 1874. Ueber eine neue Methode der maassanalytischen Bestimmung des Silbers. *J. Prakt. Chem.,* 117 : 217–224.
302. Volhard, Jakob, 1877. Die Anwendung des Schwefelcyanammoniums in der Massanalyse. *Ann. Chem. Pharmacie,* 190 : 1–61.
303. Von Arx, William S., 1962. *An Introduction to Physical Oceanography.* Addison-Wesley, Reading, Mass., 403 pp.
304. Von Meyer, E., 1906. *A History of Chemistry* (Translated by George McGowan). MacMillan, London, 655 pp.
305. Walker, J., 1928. Arrhenius memorial lecture. *J. Chem. Soc.,* 1 : 1380–1401.
306. Watson, William, 1753. An account of Mr. Appleby's process to make sea-water fresh; with some experiments therewith. *Phil. Trans. Roy. Soc. Lond.,* 48 : 69–71.
307. Weeks, M.E., 1956. *Discovery of the Elements.* Journal of Chemical Education, Easton, Pa., 6th ed., 898 pp.
308. Werner, A.G., 1962. *On the External Characters of Minerals.* (Translated by A.V. Carozzi.) University of Illinois, Urbana, 118 pp.
309. Wilkinson, John, 1765. A course of experiments to ascertain the specific buoyancy of cork in different waters: the respective weights and buoyancy of salt water and fresh water: and for determining the exact weight of human and other bodies in fluids. *Phil. Trans. Roy. Soc. Lond.,* 55 : 95–105.

310. Williams, F.L., 1963. *Matthew Fontaine Maury, Scientist of the Sea.* Rutgers University, New Brunswick, N.J., 720 pp.

311. Wilson, George, 1849. On the solubility of fluoride of calcium in water, and its relation to the occurrence of fluorine in minerals, and in recent and fossil plants and animals. *Trans. Roy. Soc. Edinb.*, 16 : 145–164.

312. Wilson, George, 1849–1850. On the presence of fluorine in the waters of the Firth of Forth, Firth of Clyde, and the German Ocean. *Chemist,* 1 : 53–54.

313. Wilson, George and Forchhammer, Georg, 1850. On the proportion of fluoride of calcium present in the Baltic, with some preliminary remarks on the prescence of fluorine in different oceanic waters. *New Phil. J. (Edinb)*, 48 : 345–350.

314. Winkler, Lajos, 1889. Die Bestimmung des im Wasser gelösten Sauerstoffes. *Mathemat. Naturwiss. Ber. Ungarn,* 6 : 176–189.

315. Winkler, Lajos, 1916. Der Jodid- und Jodat-Iogehalt des Meerwassers. *Z. Angew. Chem.*, 29 : 205–207.

316. Withering, W., 1798. Analyse chimique de l'eau de Caldas, de l'Académie de Lisbonne et de la Société Royale de Londres, *Ann. Chim., Ser.1*,25 : 180–185.

317. Wolf, A., 1959. *A History of Science, Technology and Philosophy in the 16th and 17th Centuries.* Harper, New York, 2 vol.

318. Wolf, A., 1961. *A History of Science, Technology and Philosophy in the 18th Century.* Harper, New York, 2 vol.

319. Wollaston, William H., 1820. On the discovery of potash in sea water. *Edinb. Phil. J.*, 2 : 325–326.

320. Wollaston, William, H., 1829. On the water of the Mediterranean. *Phil. Trans. Roy. Soc. Lond.*, 49 : 29–31.

321. Wooster, W.S., Lee, A.J. and Dietrich, G., 1969. Redifinition of salinity. *Limnol. Oceanogr.*, 14 : 437–438.

322. Wüst, G., 1964. The major deep-sea expeditions and research vessels 1873–1960. In: M. Sears (Editor), *Progress in Oceanography, 2.* MacMillan, New York, pp.1–52.

INDEX